Performance under Sub-Optimal Conditions

Proceedings of the symposium on

Performance under Sub-Optimal Conditions

Edited by

P. R. Davis

Department of Biological Sciences
University of Surrey
Guildford

Published by
Taylor & Francis Ltd
10–14 Macklin Street
London WC2B 5NF

1970

First published 1970 by Taylor & Francis Ltd, 10–14 Macklin Street, London WC2B 5NF.

Reprinted from *Ergonomics*, Volume 13, No. 5, 1970.

© 1970 Taylor & Francis Ltd.

All Rights Reserved. No part of this publication may be reproduced, stored in a memory system, or transmitted in any form or by any means, electronic, mechanical, photocopying, recording or otherwise, without the prior permission of the Copyright owner.

Printed in Great Britain by Taylor & Francis Ltd, 10–14 Macklin Street, London WC2B 5NF.

ISBN 0 85066 044 0

List of contributors

Acton W. I.
Institute of Sound and Vibration Research
University of Southampton
Hampshire

Aldridge J. F. L.
I.B.M. United Kingdom Ltd
London, W.1

Atherley G. R. C.
Department of Pure and Applied Physics
University of Salford
Lancashire

Bryson D. D.
Imperial Chemical Industries Ltd
Nobel Division, Stevenston
Ayrshire

Colquhoun W. P.
Applied Psychology Unit
Medical Research Council
Cambridge

Duncan Catherine
Department of Pharmacology and MRC Air Pollution Unit
St. Bartholomew's Hospital Medical College
London, E.C.1

Gibbons S. L.
Department of Pure and Applied Physics
University of Salford
Lancashire

Glover J. R.
Department of Social and Occupational Medicine
Welsh National School of Medicine
Cardiff

Guest A. D. L.
Department of Pharmacology and MRC Air Pollution Unit
St. Bartholomew's Hospital Medical College
London, E.C.1

Lawther P. J.
Department of Pharmacology and MRC Air Pollution Unit
St. Bartholomew's Hospital Medical College
London, E.C.1

Malcolm D.
Royal College of Physicians
London, N.W.1

Murray R.
Trades Union Congress
Great Russell Street
London, W.C.1

Powell J. A.
Department of Pure and Applied Physics
University of Salford
Lancashire

Rt. Hon. the Lord Robens of Woldingham
National Coal Board
London, S.W.1

Singleton W. T.
Applied Psychology Department
University of Aston in Birmingham
Birmingham 4

Smith G. F.
Department of Employment and Productivity
Clifton Down
Bristol BS8 4AT

Turvey R. J.
H. J. Heinz Co. Ltd.
London, N.W.10

Wyon D. P.
Statens Institut for Byggnadsforskning
Lund
Sweden

Contents

v   List of contributors

ix   Preface

1   Opening address
      The Rt. Hon. Lord Robens of Woldingham

5   Introduction to session 1
      W. T. Singleton

6   Moderate acoustic stimuli: the interrelation of subjective importance and certain physiological changes
      G. R. C. Atherley, S. L. Gibbons and J. A. Powell

16   Speech intelligibility in a background noise in noise-induced hearing loss
      W. I. Acton

27   Introduction to session 2
      D. Malcolm

28   Circadian rythms, mental efficiency and shift work
      W. P. Colquhoun

31   Health and productivity
      D. D. Bryson

39   The stresses and strains on feet in industry
      R. J. Turvey

49   Introduction to session 3
      J. R. Glover

50 The investigation of the mental effects of trichlorethylene
G. F. Smith

57 Carbon monoxide and phenobarbitone: a comparison of effects on auditory flutter fusion threshold and critical flicker fusion threshold
A. D. L. Guest, Catherine Duncan and P. J. Lawther

67 Introduction to session 4
R. Murray

68 Studies of children under imposed noise and heat stress
D. P. Wyon

83 Emotional illness and the working environment
J. F. L. Aldridge

93 Subject index

Performance Under Sub-Optimal Conditions

In 1967 the British Occupational Hygiene Society and the Society of Occupational Medicine joined with the Ergonomics Research Society in organizing a symposium on the effects of extreme environmental conditions on working performance. The symposium was highly successful, and as a result the three societies agreed to continue their collaboration with further symposia, and this number includes a report of the proceedings of the second two-day symposium, held on 6th–7th May 1970, at the Zoological Society of London. Its subject was ' Performance under sub-optimal conditions '.

The opening address was given by the Right Honourable the Lord Robens of Woldingham, P.C., who for most of his life has been concerned with industrial environments. Following the general introduction the meeting was held in four sessions, these being under the chairmanship of Professor W. T. Singleton (Chairman, E.R.S.), Dr. D. Malcolm (President, S.O.M.), Dr. J. R. Glover (President, B.O.H.S.) and Dr. R. Murray (T.U.C.) respectively. After each session a panel discussion of the papers was held, and brief reports of these are presented here.

Opening Address

By the Right Honourable the Lord Robens of Woldingham

It is a great privilege to have the opportunity to address you and open this two-day conference—the second of its kind which brings together members of the British Occupational Hygiene Society, the Ergonomics Research Society and the Society of Occupational Medicine.

The modern world has bought a tremendous amount of progress at the expense of increasing, and perhaps already excessive specialization. It is daily becoming more and more true that in order to attain the boundaries of knowledge and push research even further forward, men are having to specialize and restrict their effort not to groups of problems, not to individual problems, but to parts of problems.

This process is perhaps inevitable in view of our human limitations; we can cram just so much and no more knowledge and experience into our heads, so if the human race is to increase its store of knowledge this must be divided out between specialist workers.

It is precisely because of this process that conferences such as this are of so much value—for it is only by coming together, as you are today, that, as individuals, you can see clearly how your own work links in to that of specialists in related fields—and this you must do, for we are rapidly coming to the stage where it is only groups of people that can solve any problem—only groups of people because no one person has a wide enough, rather than deep enough, body of knowledge and experience to know the answer on his own.

All of you are interested in some facet of the problem of the human being at his place of work—the problem of seeing that the working environment does not unduly interfere with the job that a man is paid to do. There are two aspects to this problem. On the one hand we are concerned with the efficiency of work, to compare the value or social benefit that a man can create working in one environment rather than in another—in short, to see whether it is worth sending men down mines, or out to sea in trawlers.

On the other hand we are concerned with the individual man—with the reaction of his working conditions on his bodily, emotional and social needs, and with the wear and tear of his daily task.

The two clearly go together—and we base our work on the belief that they do—but, as we all know, this has been by no means always the case. Numerous examples of adverse working conditions abound, conditions which appalled some even of our Victorian grandfathers. As early as 1796 **Dr.** Percival's report, drawn up on behalf of the Manchester Board of Health, hit the nail squarely on the head with its second conclusion, that ' The large factories are generally injurious to the constitution of those employed in them, even where no particular diseases prevail, from the close confinement which is enjoined, from the debilitating effects of hot and impure air, and from the want of the active exercises which nature points out as essential to invigorate the system and to fit our species for the employments and the duties of manhood.'

Dr. Percival was unfortunately ahead of his time in a world in which Andrew Ure, a Victorian industrialist, could assert that:

' By the infirmity of human nature it happens that the more skilful the workman, the more self-willed and intractable he is apt to become, and, of course, the less fit a component of a mechanical system, in which by occasional irregularities he may do great damage to the whole. The grand object of the modern manufacturer is, through the union of capital and science, to reduce the task of his work-people to the exercise of vigilance and dexterity.'

Fortunately the parliamentary and public action which sprang from Dr. Percival's report overthrew the nightmare paradise of Andrew Ure, a world fit only for F. W. Taylor's trained gorillas or Aldous Huxley's Epsilon Minus's, and it is our task now to improve the conditions of the men at work, rather than merely to make them bearable.

The nature, and at the same time a major difficulty, of your task is that in human engineering more than in any other branch of engineering you are dealing with a very variable commodity.

It is true that people do have broadly similar physical and emotional needs, and will perform best when those needs are satisfied. However, these needs are not exactly similar, except in a few examples of limited, though by no means unimportant, interest. All people perform better in stable conditions—and to be hurled around on the decks of a trawler does nothing to increase the ease with which any job is done. Dust causes medical problems everywhere and to everyone, though below a certain level it may have no measurable physiological or efficiency impact. The task of the practitioner in these fields is clear—to reduce the level of disturbance by every means at his command.

Much more important though are the conditions which we *all* face to one extent or another—working hours, noise, temperature, lighting conditions, fresh air, the physical movement and effort needed to do the job, eating and resting and sleeping habits. Measured in these terms people do differ significantly and perform the same task best under comparatively widely varying conditions.

Take temperature, for instance. Although the Industrial Health Research Board concluded, as long ago as 1940, that in winter, temperatures for moderately heavy work should be between 60 and 65 degrees Fahrenheit, actual enquiries have found time and again that some will be comfortable and some will be uncomfortable whatever temperature is chosen in a range as wide as 54–76 degrees Fahrenheit.

Lighting is another problem. Again, people have widely varying preferences and it is difficult to work out how far improved performance on a change of conditions is due to direct physical influence—such as clearer vision from improved lighting—how much to individual likes or dislikes for bright or dim lighting and how much to the effect of change in itself, as evidence that an employer takes interest in his work-people.

This is very important, as it strikes to the heart of the problem we are facing. How far can we separate ' the interior decorating ' aspects of the work situation from the underlying and essential structural building work that has to be done on the work-place.

It is a chastening thought that some people perform extremely well in physical conditions which are objectively appalling—one need only think of soldiers in war or of astronauts cooped up in a capsule on the way to the moon—yet others perform badly under apparently ideal conditions, in beautiful offices where every care has been taken to suit the surroundings and the equipment to the job in hand. The clue to this problem lies inside man himself. There is something inside man, morale, a sense of purpose, a sense of achievement which can upset all the calculations we do.

This is made abundantly clear in my own industry. Over the last ten years much has been done to improve underground working conditions and the equipment for the job—and this investment has been really worth while with a productivity rise to date of something like 75%, in terms of output per manshift. Yet we find ourselves in a position where with identical geology, with identical equipment, with identical organization, and with identical groups of men in terms of age and physical fitness, the productivity of different units varies by a factor of 5 times—an enormous variation which cannot easily be explained away by geographical or historical factors, or even by ' morale ', and payment systems.

A situation such as this is forbidding for the man who is working to improve working conditions. Its message is that, try as we might to improve conditions, we are faced with, and may actually be increasing, the opportunity for very wide variation in the levels of performance we can expect from people. In short, Andrew Ure may have been unknowingly right when he commented on the problem of the skilled man ' as a less fit component of a mechanical system '. I would like to put his point the other way round. Just as a mechanical system amplifies the work which can be done by a man—so also it increases the importance of the mistakes he makes, and opens up the opportunity for him to make new mistakes. The pedestrian who does not know where he is going is not much of a problem, but the car driver is—simply because the more complicated we make life, the more rules are needed to keep everything under control and the more due the consequences of losing control.

This leads into a vicious circle, with each advance increasing the variability of performance, increasing the noise of uncontrolled and unmeasured factors through which we have to listen for the whispering voice which tells us how to improve things, with it becoming more and more difficult at each turn of the circle to see what to do next.

I am of course a layman, and it is easy for a layman to offer unusable or wrong advice, but I would like to risk my neck and put one idea to you—an idea which is not wholly new, but which I think contains the germ which may lead to a very useful thought.

We always, and quite rightly, think of man as part of a larger system. On the large scale, theologians did so in the middle ages and ecologists do so now. Though the details differ, the idea is the same in each case—man is part of, and must fit into, a longer view of the world. However, merely accepting this is of no use to anyone—the idea has to be put to use, and it is here that we in industry may have made our mistake, a mistake which is only now beginning to be corrected by industrial psychology and ergonomics workers.

We have tended to say, and Andrew Ure said it most clearly, that, the way to bring men and equipment together, creating jobs is this:

Look just at the overall job that is to be done and at the machines that are, or could be made available, for doing it. Sling them together in the most economic way, and then put men into the gaps in the middle, to do the jobs which cannot be done by machines, or to operate them. When this has been done we take a good look at the unattractive Juggernaut we have created and decide where it is necessary to alter the machinery to do the job more efficiently—if as a by-product this means making the environment better for the people trapped in the machine, then the environment should be improved—otherwise people will just have to be forced to ' fit in '.

This then is my suggestion: we should give up ' fitting people in ' and think instead of building the environment around people—this is an essential part of human engineering—and it is an essential part of your work—to encourage an attitude which leads to a situation where we don't have to ask ourselves ' how do we improve these conditions ', but can instead, by clear thinking at the outset, ask ' is it really worth changing the job in such and such a way to get better overall performance, though it means a change to the worse for the individual '. To do this we must fit things around people rather than fit people in.

It is my own hope that this may be a useful idea for you when you come together during this conference, to which I wish, because it deserves, every success.

Session 1
Introduction

By Professor W. T. Singleton

It is my pleasure to act as Chairman of the first session which opens this joint conference of our three societies. Our respective governing bodies have recognized for a long time that there is extensive overlap between the interests and objectives of the societies. It is proper that we should have our separate societies with their separate meetings, but it is equally proper that occasionally we should have joint meetings where we can learn from each other—not just on matters of data and theories but also from the differences in our attitudes and techniques—the way in which we approach our problems. Our common problems for the next two days are: noise, heat, toxic hazards, shift work and more generally the stress and strain of work in modern industry. Our first session today is on problems of noise. I do not need to emphasize to this audience the importance of this topic, not just in industry but in modern life generally. Fortunately there is growing public concern about this problem which will be reflected before long in more legislation regulating noise. This is our opportunity, but it is also, of course, a considerable responsibility. We have to be very clear that we know what we are talking about before influencing dictates which will appear on the statute book. There is already some danger that public pressure will get ahead of scientific knowledge in this field. To give just one example of our ignorance: the levels which will be fixed for permissible noise levels at work will be far less than those levels which, our young people in particular, appear to regard as necessary for enjoyment in their leisure hours.

It seems to be urgent that we find out a lot more about the effects of noise. We know something about dangerous levels but we know very little about the long-term effects or short-term effects of moderate levels. This is true for other kinds of stress such as heat, cold, poor lighting, posture and so on. We know very little at the moment about combinations of stresses. There is some theoretical evidence that might suggest that mild heat stress will counterbalance noise stress, but on common-sense grounds this would appear to be very dubious. Lord Robens has already emphasized the importance of individual differences. Once again, we know very little about the effects of personality differences, age differences, anatomical and other differences, which may well be very considerable. In the stress field, as in other ergonomics fields, we shall have to stop talking about that mythical creature the 'average worker' and base our knowledge on the different reactions of different kinds of workers.

All this suggests that at present in stress research we have got no further than an awareness of the scale of the problems which require to be dealt with.

Moderate Acoustic Stimuli: the Interrelation of Subjective Importance and Certain Physiological Changes

By G. R. C. ATHERLEY, S. L. GIBBONS and J. A. POWELL

Department of Pure and Applied Physics, University of Salford, Lancashire

A preliminary study was carried out to determine the interrelation between 'moderate' acoustical stimulation and certain physiological changes. It has been shown that 'subjective importance' of the noise was a material factor effecting changes in skin resistance. Further studies were made of the effect of whole-day exposure to aircraft noise, typewriter noise and white noise. The noises of high subjective importance, the aircraft and the typewriter, both showed measurable physiological changes, whereas that of low subjective importance (white noise) showed no significant change compared with control levels. Estimations from four subjects showed a marked decrease in 24-hour urinary 17-ketosteroid and eosinophils, and an increase in total white cell count, lymphocytes and neutrophils. It is suggested that 'moderate' noise does not appear to act as a 'conventional' stressor and it is further postulated that it may result in a characteristic syndrome which is comparable with a mild form of anxiety-depression.

1. Introduction

The purpose of our research has been to make a preliminary exploration of the interrelation between moderate acoustical stimulation, subjective ranking of this stimulation and certain physiological changes. The project was conceived because there is little conclusive evidence to confirm or deny the widely held belief that noise causes stress. For example, there has been much discussion over the past few years of the possible effects of aircraft noise on health. In 1963 the Wilson Report discounted any such effects and, in 1968, Atherley commented on the lack of relevant scientific evidence. In 1969 Abey-Wickrama and colleagues produced evidence that there is a higher psychiatric hospital admission rate among people living in the 'maximum noise area' around Heathrow Airport compared to that outside this area. 'Maximum noise area' is defined as having noise levels in excess of 100 PNdB (approximately 87 dBA) which by industrial standards would be judged moderate. Although the authors were careful to avoid the suggestion that aircraft *noise* itself can cause mental illness, it is clear that they are justified by their evidence in concluding that 'high intermittent noise levels may be a factor in increased rates of admission' to a psychiatric hospital. Whatever the direct cause of the observed disparity in admission rates, whether aircraft noise or fear of aircraft accident, it is certain that the results will add weight to the arguments of those who regard noise as being harmful to health. Thus the need is urgent for a detailed understanding of the psychological and physiological effects of noise.

2. The Present Research

The principle underlying our research in the laboratory was the customary approach of measurement of response to specific stimulus. As a first step it was necessary to explore certain aspects of both stimulus and response.

2.1. Measurement of Acoustic Stimuli

There are two ways of predicting subjective judgment of noisiness from objective acoustical measurements.

> The first is in terms of its absolute level and for particular categories of noise (sound), the agreement may be very good indeed. The noise of motor cars, for example, shows good agreement ($r_s = 0.86$) between subjective judgment and measurements in dBA (Ford, Hughes and Saunders 1970). Thus moderate sound could be described in values derived from the dBA scale.
>
> The second approach is to compare the intruding sound with the background level. This approach is exemplified in British Standard 4142: 1967, which sets out a method of rating industrial noise affecting mixed residential and industrial areas. Assessment is based not on an 'absolute' measure of sound but on the difference in dBA between the sound and the ambient acoustical level with which it is competing. Thus a moderate sound might be one whose excess level over background is 5 dBA.

In this series of tests we chose to adopt the former approach for the purposes of measuring 'moderate' acoustical stimuli, which for our needs were defined as being below 95 dBA.

2.2. Subjective Importance

It is often said that 'noise is unwanted sound' which implies a distinction at a psychological level. If such a distinction is accepted as valid, it is to be expected that the attitude of people to a *noise source* would influence their reaction to the noise. It seems possible that the factor of 'subjective importance' might operate in the perception and response to noise. Support for this view comes from Franken and Jones (1969). They have suggested that at relatively low levels, level itself is not the most important factor governing community reaction. Among the factors which they thought to have an influence is *meaning* of noise which presumably is a concept related to that of subjective importance.

At this stage it should be pointed out that a 'psychological' view is in conflict with an energy concept of the relationship between physical stimulus and biological response. For example, it has been shown recently by Robinson and Cook (1970) that the energy concept is valid in regard to hearing loss from noise. Robinson (1969) also finds evidence of an energy concept in relation to annoyance from noise. He has formulated a 'noise pollution level', L_{NP}, given by the expression

$$L_{NP} = L_{eq} + k\sigma$$

where L_{eq} is the energy mean of the noise level over a specified period expressed in dBA, σ is the standard deviation of the instantaneous level and k is a constant given as 2.56.

In the light of these and other observations, we thought it necessary to carry out a preliminary experiment to decide whether subjective importance was a material factor or whether the stimulus magnitude alone, as expressed in terms of either L_{NP} or dBA, might have been used to predict subjective judgment.

3. Experiment to Investigate Subjective Importance

3.1. Method

The purpose of this experiment was to compare the subjective importance of some sounds with the physiological response measured in terms of changes in skin resistance.

The method has been described in detail elsewhere (Gibbons 1970); here only the salient points are given. Subjects were seated in an acoustic booth and were presented, through earphones, with various tape-recorded sounds. These were: jet aircraft taxying, an alarm bell ringing, a baby crying and white noise. The average level, in dBA, the standard deviations of the instantaneous level and the derived L_{NP} values are shown in Table 1. Each sound was presented for three minutes and there was a separation between them of two minutes of silence. Skin resistance (palm-dorsum of hand) was measured continuously with a chart recorder. At the conclusion of the experiment each subject was asked to put the sounds 'in order of subjective importance'. Subjective importance was explained in a general way to the subjects but not defined closely; they seemed to have little difficulty in arriving at a rank order for the sounds.

3.2. Results and Discussion

The pattern of change in skin resistance was similar in all subjects. At the onset of each sound there was a sharp decrease in resistance (seen on the chart as a peak of the trace). Following this, the skin resistance returned towards its original level in an exponential fashion. Gibbons has shown that 'peak height' is an onset phenomenon and is not a useful guide to the overall response. He found that a valid and reliable measure is the decay time (the time t, in seconds, for the trace to fall to $1/e$ of its original value).

Results were analysed from 14 male and 14 female subjects; overall the relationship was very good between the ranking according to subjective importance and decay times ($r_s = 0.88$, $p < 0.01$). The decay times for the white noise stimulus were compared with those from other sounds, using Wilcoxon Matched-Pairs Signed-Ranks Test. For the subjects as a group the aircraft, baby and bell all showed significantly longer decay times than the white noise.

From these results it was concluded that, in the particular experiment, there was a close relationship between the subjective importance and the change in skin resistance as measured by t. It can be seen from Table 1 that the four measured values of L_{NP} did not fall in the same order as the skin resistance or subjective ranking.

Table 1. Comparison of subjective and objective rankings

Sound	L_{eq} (dBA)	SD (σ)	L_{NP}	Objective ranking (using L_{NP})	Subjective and physiological ranking
Baby	89.5	21.5	144.5	1	3
Aircraft	94.6	5.5	108.7	2	1
Bell	93.0	1.0	95.6	3	2
White noise	92.5	1.0	95.1	4	4

4. Physiological Changes

It has been shown in the previous experiment that some moderate sounds of high subjective importance produce changes, of a temporary nature, in skin resistance. Hume (1966) expresses a widely held view that such electrodermal activity is associated with tonic and phasic activity in the sympathetic division of the autonomic nervous system. The adrenal medulla is innervated by the sympathetic division and stimulation causes release of adrenaline and noradrenaline from the medulla. An early view of the sympathetic system as a whole was that activity in it mobilizes the body's resources for 'fight and flight'. However, certain recent observations (Grossman 1967, Darrow 1947, 1950) are not wholly in accord with this view and there is a possibility that sympathetic activity is more a reflection of emotion and state of arousal than a precursor of urgent physical activity. For example, it appears that noradrenaline may increase the heart rate without necessarily increasing circulation. It is known that changes in skin resistance occur not only in response to emotional stimuli but also to all sudden or unusual events or changes. Thus alterations in skin resistance should be seen as a manifestation of more general physiological changes associated with sympathetic activity. Grossman summarized the evidence in his observations that skin resistance relates quite well with phenomena such as 'arousal, general activation or attention' where these are appropriately defined. It seems to follow from Darrow's proposals that prolonged sympathetic activity is associated with feelings of anxiety.

In the present context, stress is used to imply a challenge which results in an adaptive reaction of a neuroendocrinal nature. Whether or not Selye's hypotheses (1950) are accepted, increased secretion of Adrenocorticotrophic Hormone (ACTH) from the pituitary is recognized as a fundamental part of physiological response to a number of different stresses. Thus, if noise is a stressor, we should find evidence of increased secretion of ACTH during exposure. In this connection we were puzzled by the results of a study by Sakamoto (1959) in which he showed that industrial workers exposed to noise of a moderate level excreted less than normal levels of urinary 17-ketosteroids. Admittedly, 17-ks levels are not usually considered an ideal index of adrenal cortical activity, and hence ACTH secretion, because the adrenal cortex is not their sole site of origin. However, because the confounding effect of these other sites would lead to an excessive production of urinary 17-ks, levels below normal are considered by us to be a relatively true reflection of decreased adrenal cortical activity. Because the level of activity of the adrenal cortex is a fundamental consideration in the question of whether noise acts as a stressor, it was decided to establish the direction of change in 17-ks in response to moderate stimulation by noise of high subjective importance. The experiment that was carried out is described in Section 4.1.

In addition it was decided to observe changes in cellular composition of the blood. Selye (1950), among others, described the following changes in response to stress; a decrease in lymphocytes and eosinophils, and an increase in neutrophils and total white cell count. There is known to be a relationship between the level of ACTH secretion in blood and eosinophils (intravenous administration of ACTH causes a decline in circulating eosinophils (Thorn Test)). Thus in certain respects the cellular composition of the blood is linked with a fundamental aspect of a response to stress.

4.1. Investigation of Physiological Changes

4.1.1. Method

The purpose of this experiment was to observe any changes in certain physiological measures (see below) during periods of exposure to three particular sounds: aircraft, typewriter and white noise. The white noise and aircraft both had an average level of 95 dBA and were chosen because in the previous experiment the former had the lowest rating of subjective importance and the latter had the highest, with the subjects tested. Typewriter noise was chosen as an example of recurrent impact noise. Previous studies of loudness of recurrent impact noise (Powell 1970) led us to suppose that, compared with continuous sound, the typewriter noise might have a relatively large physiological effect; for this reason the stimulus level was set low (70 dBA average level).

The seven-hour exposures took place on three separate days of a five-day randomized experiment; the two non-exposed days provided control information. The exposures were intended to reproduce a working day spent in a moderately noisy environment so that meal and tea breaks were permitted during which there was freedom from stimulus. The total exposure time amounted to seven hours per stimulus. Care was taken to ensure that the subjects' patterns of life were consistent throughout the entire experiment. The four subjects were university staff with sedentary occupations and their duties over the five-day period involved mostly reading and writing. The stimulus was delivered through earphones and was provided by a portable tape-recorder which the subjects carried with them wherever they went. On one of the control days the subjects wore the earphones and carried the tape-recorder, but were given no sound stimulus.

The following were studied; urinary 24-hour 17-ketosteroids, and two-hourly estimations of total white cell count (WCC), eosinophils, neutrophils and lymphocytes. The biochemical and haematological techniques employed have been described elsewhere (Gibbons 1970).

4.1.2. Results

17-ks

The median values for the four subjects are shown in Figure 1. It can be seen that relative to the control days there was no difference in 17-ks output during the 24 hours which included the exposure to white noise. During the period corresponding to typewriter and aircraft noise the 17-ks output was less than that for the days of either the control or the white noise.

Total white cell count, lymphocytes, eosinophils and neutrophils

Median values of two-hourly readings are shown in Figures 2–5. These are given in relation to the control level. In order to discount the effect of circadian variation the control and experimental values were matched in time. Overall there was an increase in total WCC, lymphocytes and neutrophils in relation to controls. The eosinophils showed an initial decrease followed by a subsequent increase. Of the blood counts, the least reliable is undoubtedly that of the eosinophils. The proportion of these cells is small and thus a counting error could have a substantial effect. For the rest of the counts we

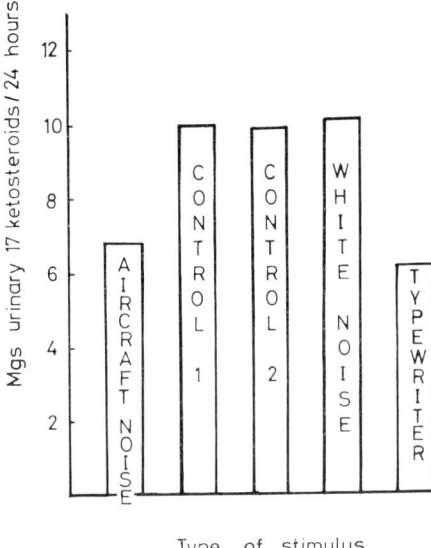

Figure 1. Median 17-ketosteroid changes.

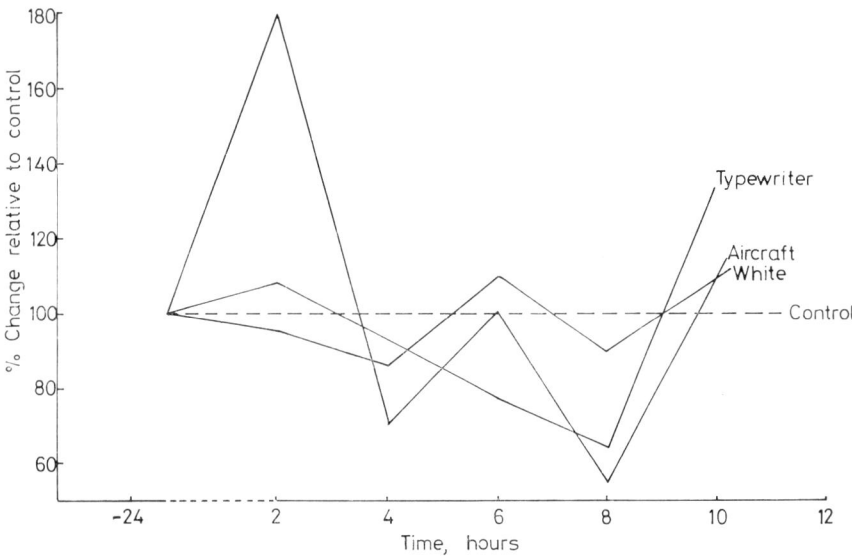

Figure 2. Median-eosinophil changes.

are satisfied that the technique adopted was adequate and that the changes observed were real ones.

Of the three stimuli used in the second experiment, aircraft noise, according to the skin resistance experiments, was of high subjective importance to the four subjects and the white noise was of low subjective importance, although the two were equal on a physical level. The third stimulus, the typewriter, although 20 dBA less in magnitude, evoked the same changes as aircraft noise. For present purposes we have assumed that typewriter noise is of high subjective importance for these four subjects. The changes observed on the

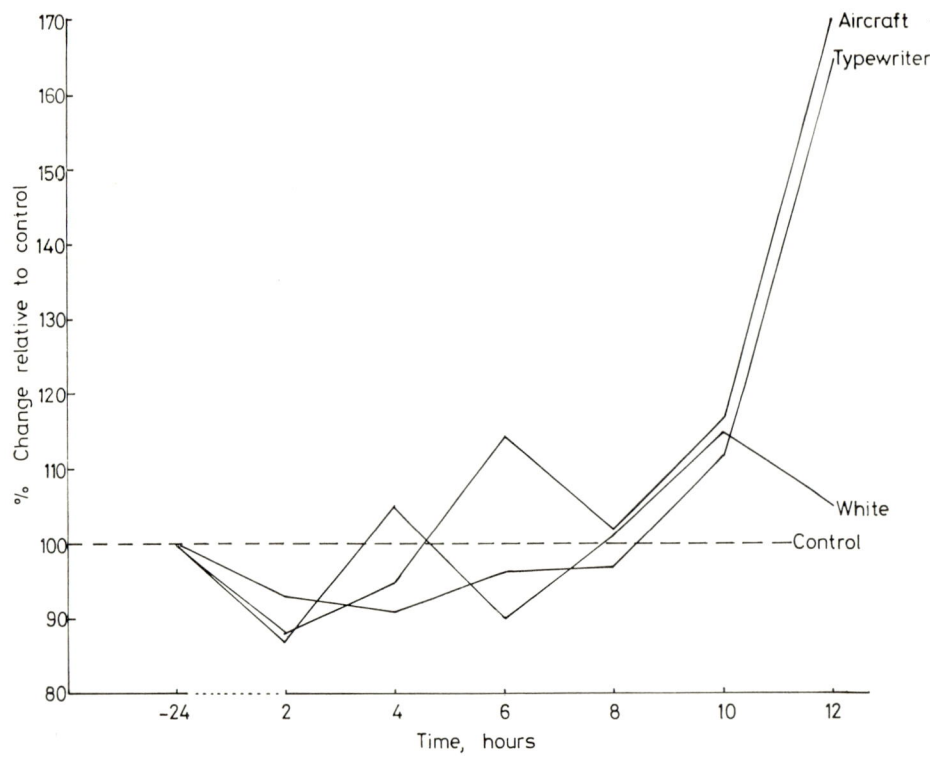

Figure 3. Median neutrophil changes.

high and the low categories are summarized in Table 2, which also shows the relevant changes which would be expected in response to a 'conventional' stressor.

Table 2. Changes observed with noise of high and low subjective importance

	Direction of physiological change (relative to control)		
Stimulus	Increase	Decrease	No change
Noise of high subjective importance	Lymphocytes Neutrophils WCC	17-ks Eosinophils (at first)	
Noise of low subjective importance			17-ks Eosinophils Neutrophils WCC
'Conventional' stressor	17-ks Neutrophils WCC	Eosinophils Lymphocytes	

It is clear that the sounds of high subjective importance evoked certain physiological changes, but these are not what would be expected from a 'conventional' stressor.

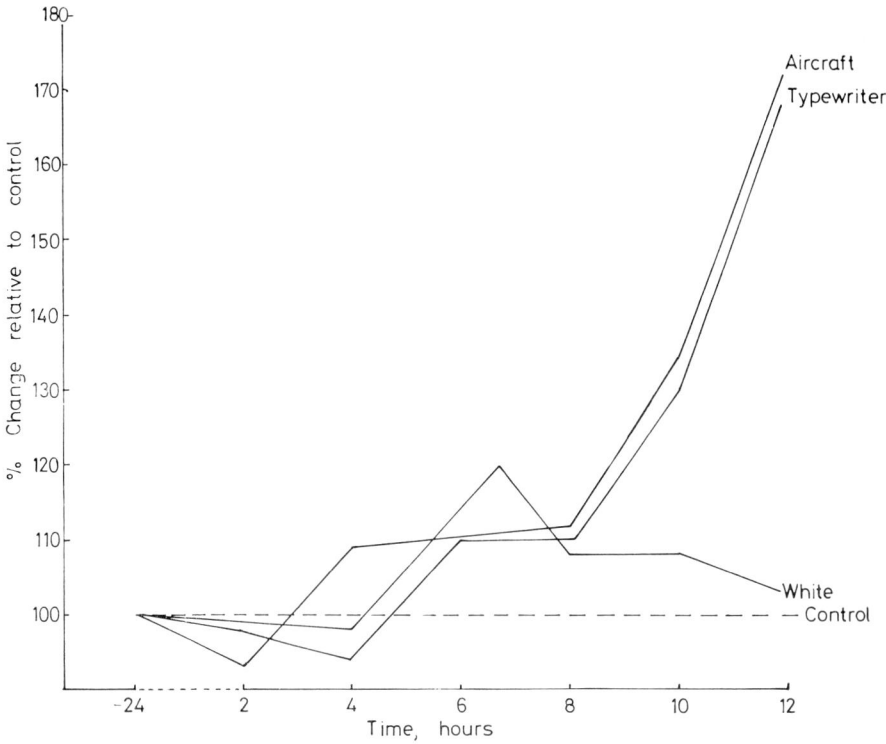

Figure 4. Median total white cell changes.

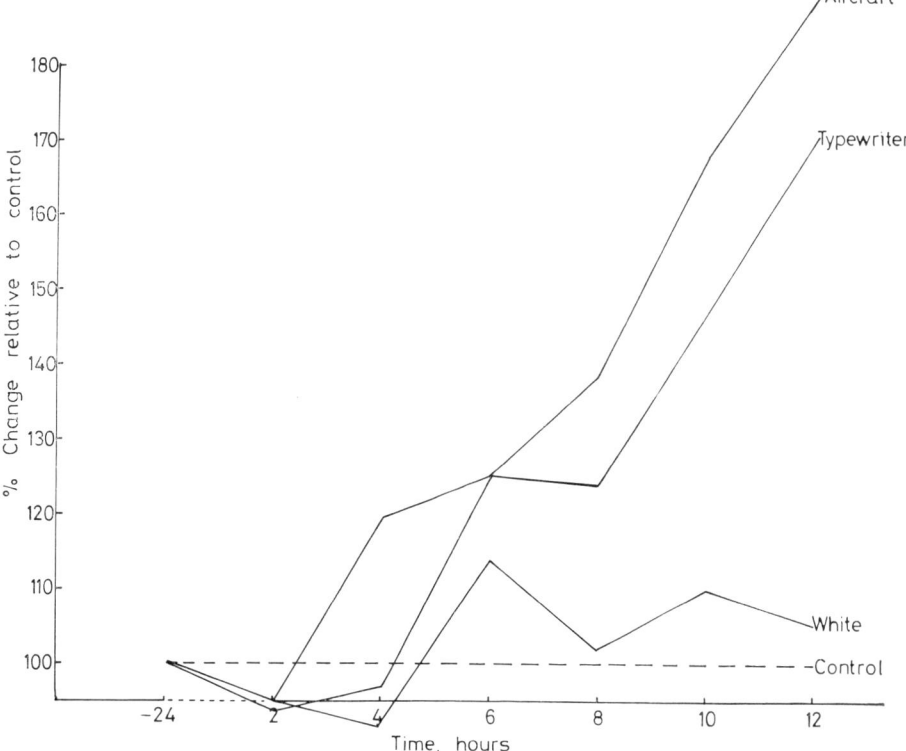

Figuer 5. Median lymphocyte counts.

4.1.3. Discussion

Overall, the findings are apparently paradoxical. On the one hand, in response to sounds of high subjective importance we observed changes in skin resistance which seem to us to be indicative of sympathetic activity. According to Levi (1967) it is well established that the sympatho-adrenomedullary system plays an important role in the general response to stress. On the other hand, we found no evidence of increased ACTH production (such as we would expect in response to stress) even with sounds of high subjective importance; the evidence, in fact, suggested the reverse.

Our work seems to suggest that noise of high subjective importance caused increased adrenal medullary activity and at the same time diminished adrenal cortical response.

Firm conclusions are not possible at this stage, but one explanation of the paradox was suggested to us by the reports of our four subjects following the days of noise exposure. After the aircraft and typewriter noises, the subjects reported feeling tired and irritable. These subjective reports are in agreement with those from a survey of noisy industries in Australia (Reilly 1959). We wonder whether the response to noise of high subjective importance ultimately results in a mild type of 'anxiety-depression' syndrome where the anxiety is associated with increased sympathetic tonus and the depression reflects diminished adrenal cortical activity. Although this is no more than speculation, it fits in with the finding of Hoagland (1957) and colleagues who have studied fluctuations in biochemical levels in relation to mental states of schizophrenic patients; when quiet and withdrawn, 17-ks secretions are low in such patients and in the aggressive phase the reverse is true.

5. Conclusions

We believe that continued exposure to noise of 'moderate' levels and of high subjective importance may lead to a characteristic syndrome associated with:

(i) subjective complaints of tiredness and irritability;

(ii) decrease in secretion of urinary 17-ketosteroid relative to the control levels;

(iii) changes in cellular composition of the blood relative to the control levels.

No such changes were observed with noise of the same acoustical intensity and of low subjective importance.

Une recherche pilote a été effectuée dans le but de mettre an évidence les effets d'une stimulation sonore modérée sur certaines variables physiologiques. Il a été démontré que l' " importance subjective " du bruit constituait un facteur matériel affectant la résistance électrique cutanée. D'autres études ont été réalisées sur les effets d'une exposition journalière au bruit d'avion, au bruit de machine à écrire et au bruit blanc. Les bruits qui présentent une grande valence subjective, tels que les bruits d'avions ou de machine à écrire, entraînent des modifications physiologiques notables, alords que les bruits de moindre valence subjective, tel que le bruit blanc, n'affectent guère ces variables.

Les estimations obtenues chez quatre sujets montrent une chute importante du taux d'excrétion par 24 heures des 17-cétostéroïdes urinaires, ainsi que des éosinophiles; en même temps l'on constate un accroissement des leucocytes, des lymphocytes et des cellules neutrophiles. On suppose que le bruit de niveau sonore moyen n'agit pas comme une contrainte banale et qu'il peut faire apparaître un syndrome caractéristique apparenté à une forme discrète de psychose dépressive.

In einer Voruntersuchung wurde die Beziehung zwischen mässiger akustischer Reizung und bestimmten physiologischen Veränderunegn ermittelt. Es wurde gezeigt, dass die subjektive Relevanz des Lärms ein realer Faktor ist, der den elektrischen Widerstand der Haut ändert. Es wurde weiter die ganztägige Wirkung von Flugzeuglärm, Schreibmaschinenlärm und weissem Lärm untersucht. Die Lärmarten mit hoher subjektiver Relevanz wie Flugzeuglärm und Schreibmaschinenlärm zeigten beide messbare physologische Veränderungen. Der weisse Lärm dagegen mit seiner geringen subjektiven Relevanz ergab keine signifikanten Änderungen im Vergleich zu Kontrollversuchen. Schätzungen an vier Versuchspersonen zeigten eine deutliche Abnahme des 24-stündigen Urin-17-Ketosteroids und der Eosinophilen, bei einer Zunahme der totalen Zahl der weissen Blutzellen, Lymphozyten und Neutrophilen. Es wird angenommen, dass " mässiger " Lärm nicht als üblicher " Stressor " wirkt, sondern eher vermutet, dass ein charakteristisches Syndrom besteht, das einer milden Form von Angst–Depression vergleichbar ist.

References

ABEY-WICKRAMA, M. F., A'BROOK, F. E., GATTON, I. G., and HERRIDGE, C. F., 1969, Mental hospital administration and aircraft noise. *The Lancet*, 13th Dec.

ATHERLEY, G. R. C., 1968, *Documenta Geigy: Noise*, (GEIGY LTD.: Manchester) 419.

B.S. 4142, 1967, *Method of Rating Industrial Noise Affecting Mixed Residential and Industrial Areas* (London: BRITISH STANDARDS INSTITUTE).

DARROW, C. W., 1947, Psychological and psychophysiological significance of the electroencephalogram. *Psychological Review*, **54,** 157–168.

DARROW, C. W., 1950, Neurophysiological effect of emotion on the brain. In *The Second International Symposium on Feelings and Emotion* (Edited by M. L. REYMERT) (New York: McGRAW-HILL).

FORD, R. D., HUGHES, G. M., and SAUNDERS, D., 1970, The measurement of noise inside cars. *Applied Acoustics*, **3,** 69–84.

FRANKEN, P. A., and JONES, G., 1969, On response to community noise. *Applied Acoustics*, **2,** 241–246.

GIBBONS, S. G., 1970, Physiological effects of noise. Ph.D. Thesis, University of Salford, Lancs.

GROSSMAN, S. P., 1967, *A Textbook of Physiological Psychology* (New York: WILEY INTERNATIONAL EDITION).

HOAGLAND, H., 1957, *Hormones, Brain Function and Behaviour* (New York: ACADEMIC PRESS).

HUME, W. F., 1966, Electrodermal measures in behavioural research. *Journal of Psychosomatic Research*, **9,** 383–391.

LEVI, L., 1967, *An Introduction to Clinical Neuroendocrinology* (Edited by E. BAYLIOZ) (Basel and New York: KARGER S.).

POWELL, J. A., 1970, The response of humans to impact noise. *Prepared for the British Acoustical Society Meeting BUILDING ACOUSTICS*, 8th–10th April 1970.

REILLY, N., 1959, The problem of noise in industry. *Medical Journal of Australia*, **1,** 2000.

ROBINSON, D. W., 1969, Concept of noise pollution. *N.P.L. Aero Report No.* AC38. March 1969.

ROBINSON, D. W., and COOK, C., 1970, Appendix 2. In *Hearing and Noise in Industry* (Edited by W. BURNS and D. W. ROBINSON) (London: H.M.S.O.).

SAKAMOTO, H., 1959, Endocrine dysfunction in noisy environment: Reports 1 and 2. *Mie Medical Journal*, ix, 39.

SELYE, H. J., 1950, *Stress* (Montreal: ACTA).

WILSON REPORT, 1963, *Committee on Problems of Noise: Final Report* (London: H.M.S.O.).

Speech Intelligibility in a Background Noise and Noise-induced Hearing Loss

By W. I. Acton

Wolfson Unit for Noise and Vibration Control, Institute of Sound and Vibration Research, University of Southampton, England

A significant improvement in speech intelligibility in a background noise was shown in a group of industrial subjects conditioned to working in noise compared with a control group of university staff. Progressive deterioration of speech intelligibility in noise was found with noise-induced hearing loss after losses had occurred at the 2 kHz pure-tone audiometric frequency.

1. Introduction

Industrial workers exposed to noise for the first time, or even exposed to a noise of different frequency spectrum, undergo a conditioning process and often speak of 'getting used to the noise'. Informal observation suggests that this conditioning process may develop over a period of two to four weeks, during which time workers are aware of a subjective improvement in their speech intelligibility against background noise. However, 'getting used to the noise' does not confer immunity to incurring noise-induced hearing loss (NIHL) after continued noise exposure over a longer period, and it is well known that older workers in certain industries suffer a diminution in their understanding of speech.

In situations where personnel depend upon speech communication, any interference with speech intelligibility may have potentially serious consequences. As pure-tone audiograms are the most readily obtainable measure of hearing, an objective relationship between pure-tone hearing levels and speech intelligibility in a background noise is of value in assessing the possibility of a failure in speech communication due to NIHL. In this investigation, the speech intelligibility scores obtained against background noise for 27 industrial subjects, divided into three groups with varying degrees of NIHL, were compared with the results from a group of 9 control subjects. A statistically significant conditioning factor was demonstrated in the group with the less severe hearing loss, and the noise-masked speech intelligibility curves have been related to the pure-tone audiograms in the two groups with the more severe hearing losses.

2. Test Conditions

As a prerequisite of realistic listening conditions, the subject must be able to make use of the advantages to be gained by listening with both ears. One advantage to be gained from binaural hearing is the ability to 'squelch' reverberation and background noise as compared to a system with a single channel feeding both ears simultaneously and equally, for example, earphones connected by a 'Y' lead (Koenig 1950). This should not be confused with directional hearing or localization, or with head shadow effects where the speech and interfering noise come from widely separated directions.

In everyday listening in semi-reverberant conditions, the background noise frequently comes from more than one source, and with the listener facing the signal source both ears hear with nore or less the same signal-to-noise ratio. The only binaural advantage which can then be obtained is by ' cross-correlation ' of the stimuli which arrive at the two ears with a slight, and variable, time delay with respect to one another due to head position and minor head movements (Cherry and Sayers 1956). When cross-correlation cannot occur, for example with identical signals arriving at each ear simultaneously or when the signals are totally dissimilar, then no advantage should be gained by the use of the second ear provided that the sensation levels at all frequencies are the same in both ears.

In order to satisfy purely physical criteria, it might seem desirable to perform any experiment under anechoic conditions, although it would have been necessary to employ a separate loudspeaker for each direction to achieve multidirectional noise. The other means of producing such test conditions is to employ a much lesser number of noise sources in semi-reverberant conditions. This is essentially a realistic everyday situation which should not prove psychologically disturbing to the subject. Furthermore, semi-reverberant rooms are readily available, and the level of background noise usually encountered in laboratories and offices does not impose serious limitations on experiments in which noise is deliberately introduced.

It was felt that the advantages of semi-reverberant conditions far outweighed those of anechoic, especially as this kept the experiment both simple and realistic. The simplest array of signal and noise sources to satisfy the experimental requirements in semi-reverberant conditions is the symmetrical arrangement of three loudspeakers shown in Figure 1.

There are three ways in which binaural hearing may occur using the apparatus shown in Figure 1, the first two mentioned being more important. Firstly, a

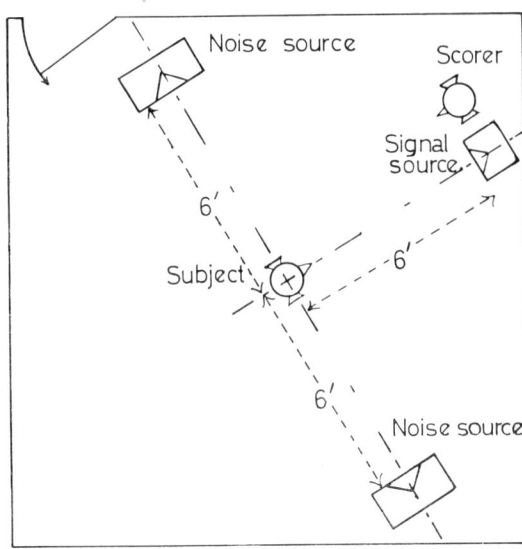

Figure 1. Experimental arrangement of loudspeakers in semi-reverberant room

certain amount of noise will arrive at each ear from the loudspeaker situated on the opposite side of the head. Secondly, the direct signals from the two loudspeakers will probably arrive at different moments in time due to the head not being positioned equidistant from each loudspeaker; a displacement of the head by 1 in. from centre gives a difference in path-lengths of 2 in. and represents an inter-aural time delay of approximately 0·2 msec. Konig (1964), on reviewing the available literature, concluded that the maximum gain in intelligibility occurs with a time delay between ears of approximately 0·7 to 0·8 msec. Finally, in semi-reverberant conditions, reflected noise will also inevitably arrive at a different moment in time with respect to direct noise, although the delay is probably too long even for first reflections to contribute significantly to binaural listening.

A separate experiment using the same speech material and group of control subjects showed a significant binaural advantage of 1·5 dB S/N ratio with this apparatus. This is somewhat less than the 3 dB binaural squelch advantage which has been demonstrated in other experiments in acoustically dead conditions (Carhart 1965, MacKeith and Coles 1970).

3. Apparatus

Three loudspeakers were arranged as shown in Figure 1 in a room measuring 14 ft by 14 ft by 8 ft 3 in. high sparsely furnished as an office. The reverberation time of this room was measured as approximately 0·3 sec using a gun-shot as the noise source. The subject was seated in a chair with a head-rest to prevent gross head movements, and a seat adjustable in height. The chair was fixed to the floor to overcome a tendency which existed for some subjects to move the chair towards the signal source as listening conditions were made more difficult. The distance between each loudspeaker and the centre of the subjects' head position, with the ears on the level of the central axis of the loudspeakers, was 6 ft.

The speech material was tape-recorded with a calibration signal at the beginning of the tape, as this enabled the same level to be presented to each subject and also removed the variable and uncertain advantage to be gained from lip-reading. The output from the tape-recorder was continously monitored by a V.U. meter, controlled by a stepped attenuator and amplified before being reproduced from the loudspeaker facing the subject. The overall frequency response of the speech channel was checked acoustically at the position occupied by the subject's head using tape-recorded random noise and a one-third-octave band filter in the measuring system, and found to be within ± 3 dB over the frequency range 125 Hz to 5 kHz, and to within ± 5 dB over the range 80 Hz to 8 kHz. In view of the semi-reverberant conditions, this was considered adequate for speech reproduction and subjective testing.

The background noise source was a 'pink' noise (i.e. random noise with constant energy per relative band-width, as opposed to 'white' noise which has constant energy per unit band-width) generator passed through a filter with a frequency response of 6 dB per octave falling towards higher frequencies. The output was controlled by a stepped attenuator, amplified and reproduced by the two lateral loudspeakers in parallel. The octave-band frequency spectrum of the background noise at the position of the subjects' head is shown in Figure 2 at an overall level of 60 dBA. Except perhaps in the higher octave bands, where the level is low, this noise has a frequency spectrum typical of many industrial noise sources.

The background noise level was set so that the control subjects gave a score of approximately 50 per cent of words correctly understood at normal voice level, i.e. approximately 65 dBA peak level at the ear for face-to-face conversation. It was found that a noise level of 60 dBA fulfilled this condition, and this level also proved satisfactory for a number of reasons of secondary importance. Masking for speech is linear within the range of noise levels from 50 dB to 90 dB sensation level (Hawkins and Stevens 1950). At the signal levels which would have been necessitated by a high background noise level, distortion occurs in the ear and leads to a decrease in intelligibility of speech. Any possibility of temporary threshold shifts occurring during the course of the experiment was removed. Finally, spoken responses from the subject back to the scorer were possible.

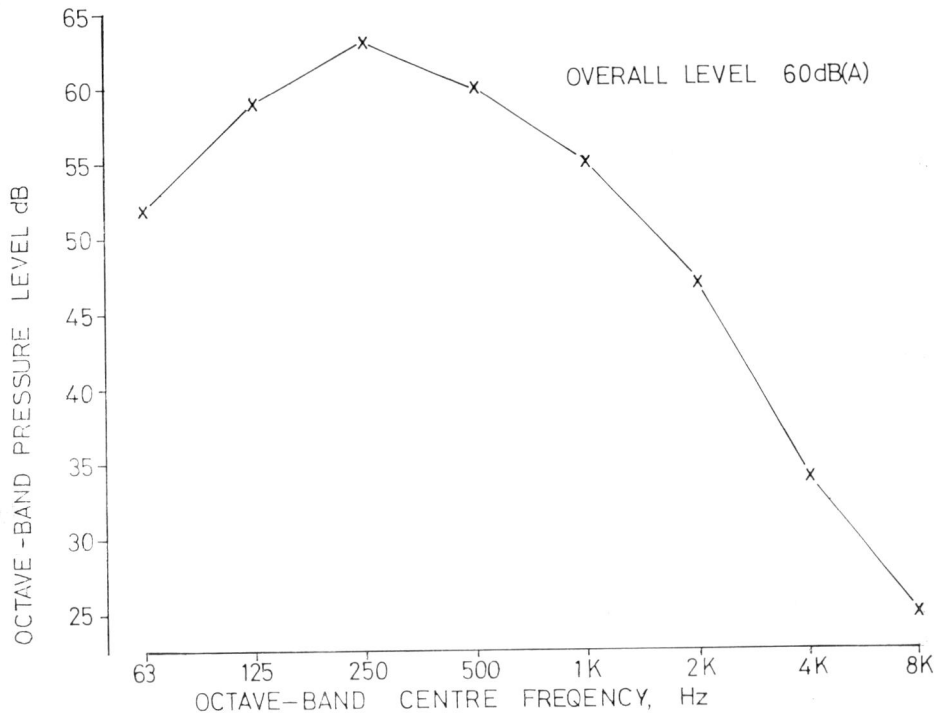

Figure 2. Octave-band analysis of masking noise

The results have been expressed in terms of signal-to-noise (S/N) ratios. The signal level was measured as the level of pink noise which gave the same deflection on the V.U. meter as the peak level of the strongest syllables, and noise level as the average reading on the 'A scale' of a sound level meter set to slow response.

4. Speech Material

Professor D. B. Fry's phonetically-balanced (P-B) monosyllabic word lists were chosen because they seem to be the most frequently used and readily available speech material in this country at the present time. Each of ten word lists is composed of 35 monosyllabic words, chosen so that each list is constructed from the same 100 phonemes, and these phonemes are representative of the distribution of phonemes in current everyday English conversational speech (Fry 1961). The words are spoken with equal effort rather than either naturally or with equal power (as measured on a V.U. meter) to allow for the wide difference in the phonetic power of various speech sounds. Also, being spoken with a 'B.B.C. accent', they overcome, to a great extent, variations in local dialects and accents, as all subjects were used to watching television or listening to the radio. Comparison of the results from the control group with the results from a matched group of subjects native to Dundee failed to show a statistically significant difference.

Each word is meaningful in itself to the subject, and there is a minimum of redundant material, although words can sometimes be completed by guessing missed phonemes. A separate experiment showed that, after an initial practice list, improvement due to learning (i.e. becoming practised at listening to this type of material as opposed to learning the word sequence) was not significant.

Although there are significant differences in intelligibility between word lists and correction factors have been derived, each word list was used at a specific S/N ratio only. As the results were compared with the results from the control group, application of these correction factors was rendered unnecessary.

5. Selection of Subjects

All subjects were put through a clinical screening procedure, and the decision to accept or reject was taken on the results of this screening in conjunction with their pure-tone audiograms. The pure-tone audiograms were obtained with a Peters model SPD5 audiometer calibrated to British Standard 2497 (1954) using the method of limits described by Hinchcliffe and Littler (1958), the mean of the descending and ascending thresholds being taken as the subjects true threshold. The hearing levels were tested at standard audiometric frequencies in the range 250 Hz to 10 Hkz inclusive. The octave-band pressure levels in the test environment were below the maximum levels allowable if the background noise in the corresponding octave band was not to mask test tones of -10 dB re threshold (Taylor $et\ al$ 1964).

Subjects were not excluded on the grounds of having acquired any part of their hearing losses due to non-industrial noise exposure, provided that they were conditioned to working in a noisy environment, and their pure-tone audiograms showed essentially symmetrical bilateral losses (subjects with between-ear differences of more than 15 dB averaged over the frequencies 500 Hz, 1kHz, 2 kHz and 3 kHz were rejected). As speech tests were involved, the further criterion was applied that all subjects must speak English as their native language.

The subjects were all employed on process plant operation or maintenance, but had been employed previously as boilermakers or shipwrights. As their present employer enforces a strict hearing conservation programme, it seems reasonable to assume that they had acquired their hearing losses in previous employment, and that they were not suffering from any temporary threshold shifts at the time of testing.

The control group were 9 young university staff (age range approximately 20 to 30 years), and only subjects with hearing levels less than 15 dB at any of the six audiometric frequencies in the range 500 Hz to 6 kHz inclusive were accepted.

6. Test Method

After audiometric screening the subjects were instructed in the procedure used for the speech tests. Instruction of the subjects played an important part in determining performance in the speech tests, especially in the case of older subjects.

Subjects were first told from which loudspeakers the signal and noise would be reproduced, as localization could have been a source of indecision on the part of the subject and hence of variance in the results. Specific instruction was given to repeat each word, irrespective of whether the whole word had been heard or not, and to guess at, or even to make an appropriate noise, for any word not fully understood. Nevertheless, it was found that some subjects tended not to respond, even though when pressed they could often correctly repeat one or two phonemes, if not the complete word. The value of using the first four word lists as initial practice lists becomes apparent, as they could be interrupted and the subject re-instructed without affecting results.

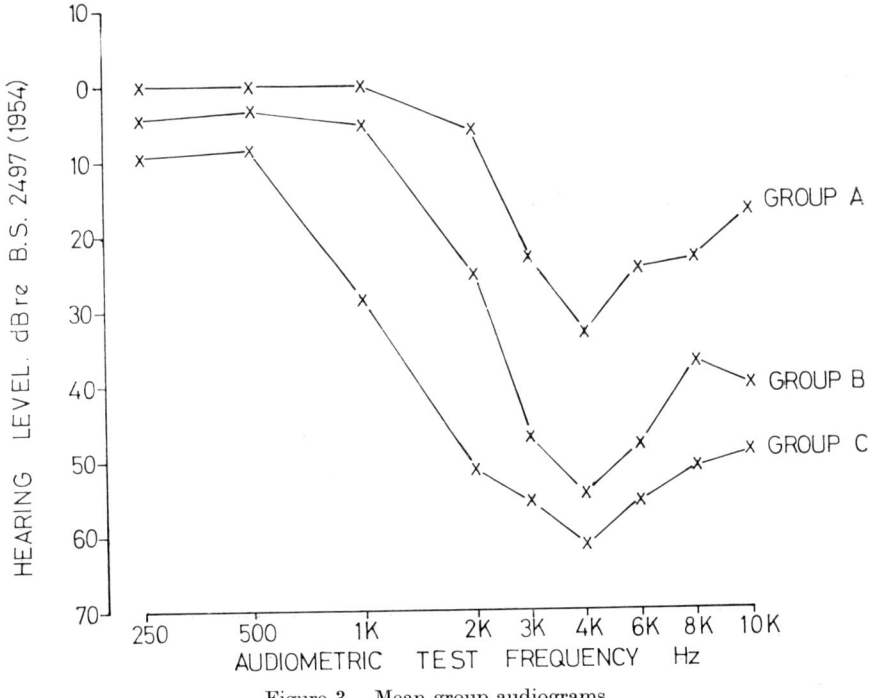

Figure 3. Mean group audiograms

Each word list was used at a given S/N ratio only, starting with word list number five at S/N ratio 20 dB. Subsequent word lists were presented at decreasing S/N ratios in 5 dB steps until word list number ten was reached at S/N ratio −5 dB, or until the subject no longer responded.

At the noise level used (60 dBA), face-to-face communication between the subject and scorer was found to be satisfactory. The scorer marked the subjects' mistakes on duplicated copies of the word lists, which were scored later by phonemes correctly repeated out of a possible total of 100. The results could thus be expressed directly as a percentage.

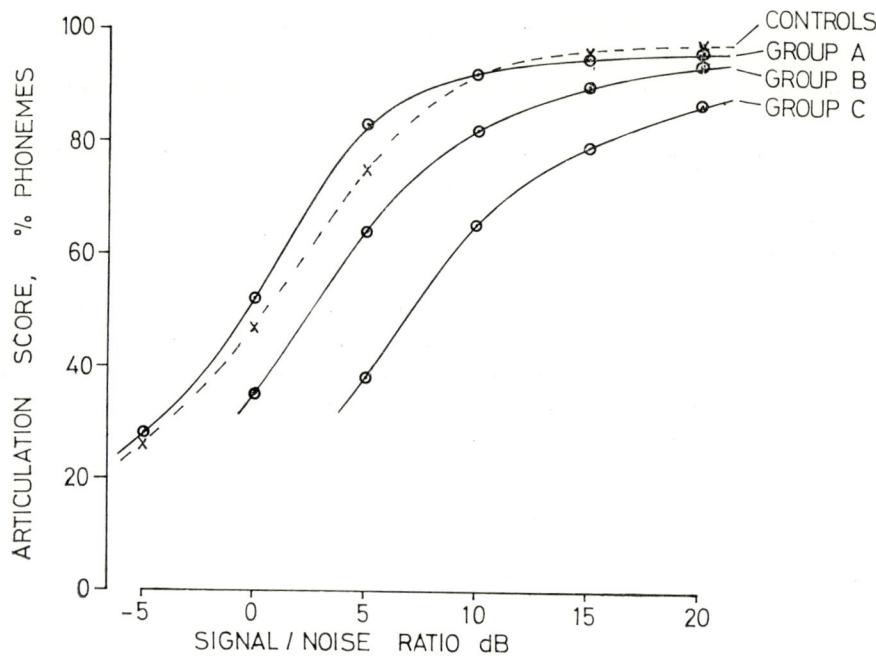

Figure 4. Comparison of articulation curves with control subjects

7. Results

The results obtained from the NIHL group were divided into three groups irrespective of the pure-tone audiograms according to the lowest S/N ratio at which a response had been obtained from the subject in the speech test. This division gave 7 subjects in the group who responded down to S/N −5 dB (Group A), 15 subjects having responded down to S/N 0 dB (Group B) and the remaining 5 subjects who had responded down to S/N 5 dB (Group C). The mean pure tone audiograms of each group are shown in Figure 3. The average ages and age ranges were 41·3, range 34 to 53 for Group A, 44·3, range 33 to 63 for Group B and 51·2, range 36 to 63 for Group C. Although the mean ages of the groups fall within a relatively narrow range, there is a rank order difference with the older subjects showing the greater speech intelligibility and pure-tone hearing losses. The control subjects all responded down to S/N ratio —5 dB.

The mean intelligibility curves for each group have been plotted in Figure 4, the percentage of phonemes correctly understood being shown as a function of signal-to-noise ratio. Inspection of this figure shows that Group A achieved, at three S/N ratios, slightly better sources than the control group, while Groups B and C respectively showed increasing divergences from the controls. Differences at higher S/N ratios were less, as the overall signal level becomes the overiding factor unless the subjects have hearing losses severe enough to cause a loss in speech discrimination even at high signal levels in the quiet.

The values of the mean divergences from the controls of each group have been presented in Table 1. The level of significance of the divergence, as shown by a t-test, has been indicated by one, two or three asterisks corresponding to probabilities of 0·05, 0·01, and 0·001 respectively.

Table 1. Deviation from controls shown by NIHL subjects (as % phonemes)

	\multicolumn{6}{c}{S/N ratio, dB}					
	−5	0	5	10	15	20
Group A	−2	−5	−8*	0	1	1
Group B		12**	11**	10**	6**	3
Group C			37***	27***	17***	10***

* $p < 0.05$ ** $p < 0.01$ *** $p < 0.001$

8. Conclusions and Discussion

Perhaps the most surprising result of this experiment was the negative divergences from the controls for Group A, which was significant ($p < 0.05$) at S/N ratio 5 dB. This group had a moderate noise-induced hearing loss involving audiometric test frequencies of 3k Hz and higher as shown by the mean group audiogram in Figure 3. The fact that the control group was comprised of university staff, who were presumably of higher I.Q. than the test subjects, and who were also familiar with listening to speech audiometric material, makes this divergence even more significant. The result suggests that the conditioning process on initial noise exposure, or ' getting used to the noise ', is, in fact, a significant factor in speech intelligibility against a background noise. More positive proof could only be obtained by a prospective study on a group of ' new starters ' in a noisy environment, and this presents immediate and obvious experimental difficulties.

Groups B and C, which showed significant losses in speech intelligibility, also showed the greater pure-tone hearing losses. Comparison of Figures 3 and 4 indicates that a significant loss, albeit slight in real terms, does not occur until the pure-tone hearing loss has progressed to involve the 2kHz audiometric test frequency, as in Group B, where the mean group hearing level was 25·3 dB at this frequency. Involvement of the 1k Hz pure-tone audiometric test frequency, as in Group C which had a mean group hearing level of 28·5 dB at this frequency, causes a highly significant and very real loss in speech intelligibility, equivalent to a S/N ratio of 6 or 7 dB. Although there was a rank-order age difference between the groups, this is not thought to have contributed significantly to the loss in speech intelligibility as the loss over this age range among otherwise normally hearing subjects was found to be less than the equivalent of a S/N ratio of 1 dB in another investigation.

The latter result may possibly be explained as follows. Redundancy present in the spoken word is progressively removed by increasing noise-induced hearing loss, but the subjects are still able to get-by up to a critical hearing loss, which will be unique for each particular subject. The mean value of this critical hearing loss presumably lies between the mean pure-tone hearing levels of Groups A and B. Removal of further redundancy by the introduction of noise masking then causes increasing difficulty of speech recognition, which bears some relationship, although perhaps not direct, with increasing pure-tone hearing loss.

The author wishes to thank Dr. R. R. A. Coles for constructive criticism throughout this work, Dr. A. Ward Gardener for assistance with the selection and screening of subjects, The Esso Petroleum Company Limited, and the subjects who cannot be mentioned by name. Financial support to the Institute from the Medical Research Council is gratefully acknowledged.

Une amélioration significative de l'intelligibilité phonétique sous l'effet d'un bruit de fond a été mise en évidence dans un groupe de travailleurs habitués à travailler dans des conditions bruyantes en comparaison avec un groupe de référence composé de personnel académique.
Une détérioration progressive de l'intelligibilité phonétique dans le bruit s'est installée consécutivement à la surdité induite par le bruit après que des pertes d'audition ait été constatées dans la fréquence audiométrique des sons purs de 2 kHz.

Das Sprachverständnis einer Gruppe von Industriearbeitern bei einem Hintergrundlärm verbesserte sich signifikant, während sich die Arbeiter an Arbeiten im Lärm gewöhnten, verglichen mit einer Kontrollgruppe von Universitätspersonal. Eine progressive Verschlechterung des Sprachverständnisses im Lärm fand sich bei lärmbedingtem Hörverlust, wenn dieser Verlust durch eine 2 kHZ-Rein-Ton audiometrische Frequenz erfolgt war.

References

BRITISH STANDARD 2497, 1954, *The Normal Threshold of Hearing for Pure Tones by Earphone Listening* (London: BRITISH STANDARDS INSTITUTION).
CARHART, R, 1965, Monaural and binaural discrimination against competing sentences. *International Audiology*, **4**, 5–10.
CHERRY, E. C., and SAYERS, B. M. A., 1956, Human ' cross-correlator'—A technique for measuring certain parameters of speech perception. *Journal of the Acoustical Society of America*, **28**, 889–895.
FRY, D. B., 1961, Word and sentence lists for use in speech audiometry. *Lancet*, p. 197–199, 22nd July.
HAWKINS, J. E., and STEVENS, S. S., 1950, The masking of pure tones and speech by white noise. *Journal of the Acoustical Society of America*, **22**, 6–13.
HINCHCLIFFE, R., and LITTLER, T. S., 1958, Methodology of air conduction audiometry for hearing survey. *Annals of Occupational Hygiene*, **1**, 114–128.
KOENIG, W., 1950, Subjective effects in binaural hearing. *Journal of the Acoustical Society of America*, **22**, 61–62.
KONIG, E., 1964, The auditory perception of space and its clinical significance. *International Audiology* **3**, 54.
MACKEITH, N. W., and COLES, R. R. A., 1970, Binural advantages in hearing of speech in noise. *Personal communication from paper in preparation*.
TAYLOR, W., BURNS, W., and MAIR, A., 1964, A mobile unit for the assessment of hearing. *Annals of Occupational Hygiene* **7**, 343–352.

Session 1

Panel Discussion

P. Sutton, Esso Petroleum. Dr. Atherley obtained his response to noise stress in the form of complaints. Complaints of this nature seem to bear a relationship to social class. Is there any evidence that lower class complainants suffer less physiological or psychological stress than those in the upper classes when exposed to similar levels of noise? Or does the lack of complaint represent a bottling-up of stress?

Atherley. Groups 3 and 4 produced their fair quota of complaints, and middle class complaints did not predominate in the study, although in other parts of the country this might not happen, as there may be regional variation. Manchester has an attenuated stockbroker belt containing a sand quarry, and complaints of noise from this source emanated from both stockbrokers and farm labourers. There is a socio-economic factor in response to noise, of course, and this is recognized in B.S. 4142, which applies different standards to different circumstances.

J. R. Glover, Welsh National School of Medicine. Will Mr. Acton be repeating his experiments using noise levels less than and greater than 60 dBA, which seems a very moderate level to have chosen.

Acton. We thought of this very early on in the experiment. We tried very low levels of noise and obtained a very wide scatter in the results. 60 dBA was chosen with great care for many reasons. If one increases the noise level, the signal level has to be increased and masking becomes non-linear; distortion occurs in the ear, and normal subjects start to show a loss of discrimination.

F. J. C. Roe, Chester Beatty. To what extent have laboratory animals been used in this type of work? Our experience showed that noise interferes with animal breeding, but that after a while this effect wears off even though the noise continues. Animals offer the opportunity to control some of the variables in work of this nature.

Atherley. Experiments have been carried out by others, but in my view it is difficult to esparate fright from noise annoyance in many of the experimental reports. However, there is nothing fundamentally wrong with using animals.

Acton. In America considerable work has been done on the relationships between noise and physiological responses, using high levels of noise over short periods. This was found to cause damage to the cochlea in animals sacrificed soon after the experiments.

R. A. Shawyer, Whitbreads and R.S.P.A. Is there any correlation between noise levels and accident proneness in factory situations?

Atherley. Froggett and Smiley showed that people vary at different times in their proneness to accidents, and that this is not inborn. The tendency to accidents is increased in parallel with other stresses and strains in people's lives: which is the cause and which the effect was not clear. If one views noise as a possible cause of stress or of anxiety, one can speculate that a man already under psychological stress may be adversely affected by extra noise.

Boyson, I.C.I. Atherley's noises were all of equal signal strength, and appear to have been unpleasant. Were any tests performed using noises of pleasant connotation, such as a revving motor cycle for an enthusiastic motor cyclist?

Atherley. No, but some of our 28 subjects had an interest in flying, and take-off noise may well have been pleasing to them. We also noticed that some subjects heard noises in the white noise, and some misinterpreted the noises we had recorded.

D. P. Wyon, S.I.B., Sweden. There was a Swedish experiment performed a few years ago in which subjects were exposed to a single noise, but shown slides of different possible sources, such as a bus, a motor cycle, a dust cart, etc. The subjective noise rating bore more relationship to the apparent source than the real noise level, and reflected the usefulness of the source to society.

Dr. Atherley's indications have great promise. Does he think that certain increased levels of excretion of indicators can be used to indicate that the noise level is becoming harmful?

Atherley. I am very hesitant about speculating on this. In practice physiological responses may help us to set standards of effect, but someone else must judge whether these effects are desirable or otherwise. A study of the physiological response can only be academic at this time.

D. Malcolm, Society of Occupational Medicine. Change in biological response does not necessarily imply damage to individual well-being. The judgment of the desirability of effects is a clinical one, and requires an assessment of the effect on the individual as an entity rather than on one small part of his physiology.

M. L. Newhouse, T.U.C. Centenary Institute. Workers in noisy environments may communicate by speaking quietly, below the background noise level. I wonder how this correlates with Mr. Acton's findings?

Acton. We were able to demonstrate effective communication at a -5 dB signal-to-noise ratio. Lip reading may play a large part, as may the content of the message. With familiar messages, only 20 per cent or so of phonemes need be correct for the message to be understood. In our experiments the phonemes were unfamiliar and lip reading was eliminated.

J. F. Eustace. Response to a stimulus depends upon the circumstances surrounding the individual at the time of exposure, and this is the most important aspect when considering the effects of noise on people.

Session 2

Introduction

By D. Malcolm

I am pleased and honoured to be asked to take the Chair at this afternoon's session of this joint meeting with the BOHS, SOM, and Ergonomics Society on Performance Under Sub-Optimal Conditions.

The most sub-optimal conditions I remember were as a student. Pathology lectures with some 200 students in a rather small unventilated lecture theatre with blinds down, lights out, slides on: we were mostly soon lulled to sleep by a soothing monotone.

As Lord Robens commented, performance is different according to motivation and many of us must have performed very well under physically much worse conditions; for instance, during the war.

There should be little danger of sleep this afternoon in this well-ventilated theatre in spite of the excellent lunch.

We start with circadian rhythms, a most important subject. With the high cost of mechanization and automation in order to obtain an adequate return on expenditure, plant must often be run continuously and shift work is becoming more common in many industries. The economic benefits must be weighed against the human and social disadvantages. Dr. Bryson has carried out a most interesting piece of research on the performance and measurement of ill health. Here is a doctor working in the field who has used his records in a most imaginative piece of work. How many more of us working in factories and offices have useful material stored up in our records?

Finally, Mr. Turner is a chiropodist who teaches at the London Foot Hospital and has, so to speak, a foot in industry with the 57 varieties.

Professor R. E. Lane in my own company started a chiropody service at the beginning of the last war when many older people were coming into industry with very bad feet and thought that their lameness was just a normal part of the ageing process. They were soon running around the factory with happy feet.

Circadian Rythms, Mental Efficiency and Shift Work

By W. P. Colquhoun

Medical Research Council, Applied Psychology Unit, Cambridge, England

1. Summary of Presentation

Shift work can be considered as one of a number of factors which might make for sub-optimal performance of the operators in any given industrial system. Detrimental effects on human efficiency which might ensue from the installation of a shift-working arrangement in such a system could arise from at least four sources. The first of these would be a lowered state of physical or mental health in the workers (though the medical evidence for this is conflicting; see Taylor 1967). The second would be the effect on motivation of having to work at nights and on weekends (times which are socially unpopular; see Wedderburn 1967). Thirdly, there is the known vulnerability of efficiency at certain kinds of task to total or even partial sleep-deprivation (Wilkinson 1965, Wilkinson et al. 1966); such a state may well occur when changing shifts, especially when going on to 'nights' and/or in rapidly rotating shift systems. Finally, there is the fact that human efficiency, particularly at mental tasks, is something which itself varies in a systematic manner, and to a not inconsiderable extent, according to the time of day or night. Thus as industrial plants become increasingly automated, and a growing number of operator functions become 'mental' in character, work will almost inevitably be carried out less effectively on certain shifts than on others, regardless of whether the first three factors are influencing the situation or not. This paper is concerned specifically with this last point, i.e. with these 'circadian', or '24-hour', rhythms in performance, and the (theoretical) possibility of reducing their effects on system efficiency by particular manipulations of duty schedules.

Unlike 'health' or 'motivational' factors, 'time of day' effects can, and have been, studied experimentally in the laboratory. It has been shown in these experiments that mental efficiency follows a distinct cycle (with a period of 24 hours), and is not simply a two-state process with 'high' or 'low' levels corresponding to the sleep/awake division of the normal day. It has also been shown that this cycle appears to follow reasonably closely the 24-hour rhythm of body-temperature.

Although most people are aware that the temperature of the body is lowered at night when sleep occurs, it is not often appreciated that it continues to fall until about 4 or 5 a.m. and that it does not subsequently reach its so-called 'normal' level during the day until about 8 o'clock in the *evening* (Blake 1967 a); thus the problem is not simply one of generally lowered efficiency on 'night' shifts as compared with others, but rather one of continuously changing levels of performance during *all* parts of the 24-hour cycle, including those sections worked by 'day' and 'afternoon' shifts. Results of studies of performance during the ordinary 'waking' day are therefore just as relevant to the present problem as those obtained from investigations of efficiency 'round the clock'.

'Waking day' studies

A remarkably wide range of mental functions has been shown to exhibit systematic, temperature-related variation during the waking day. Many of the earlier studies were summarized by Kleitman (1963). Kleitman's own work was particularly concerned with reaction time, but he also observed periodicity in card-dealing, card-sorting, 'mirror-drawing', nonsense-syllable copying, code transcription, and multiplication. A more recent study by Blake (1967 b) demonstrated that 'time of day' effects exist also in letter-cancellation, addition, prolonged attention (vigilance), serial choice reaction-time, and short-term memory (digit-span). Although, in general, Blake's performance curves followed, as did Kleitman's, the temperature rhythm, a 'post-lunch' dip was also observed with most tasks, and, in the case of short-term memory, the performance–temperature correlation appeared to be *negative*. Other workers have observed diurnal fluctuation in activities ranging from the most 'primitive' processes such as the electrical sensitivity of the eye (see Kleitman *op. cit.*) to relatively 'high-level' functions like time estimation (Thor 1962, Pfaff 1968) and responses to a word-association test (Aarons 1968). Sleep-deprivation has been shown to increase the magnitude of these 'waking-day' fluctuations in mental efficiency (Loveland and Williams 1963; Fiorica *et al.* 1968; Drucker *et al.* 1969).

'Round-the-clock' studies

These comprise investigations in which some at least of the measurements of performance have been taken in 'night' hours, i.e. at times when the subject would otherwise have been sleeping. Many of these studies have been concerned with abnormal work–rest routines in military situations; thus Kleitman and Jackson (1950) found that the speed with which 600 colours could be named was closely related to the body-temperature rhythm throughout the 24-hour period in a Naval 'split-shift' rapidly rotating watch-keeping system, and Chiles *et al.* (1968) observed a similar correlation for vigilance, arithmetic computation, problem-solving and pattern recognition during prolonged trials of aerospace duty–schedule systems with exceptionally high work–rest ratios. In other laboratory experiments 24-hour rhythmicity has been found to be present in a considerable number of measurable activities, including not only those also studied solely in 'waking' hours but, additionally, in tasks of, e.g. sensorimotor co-ordination (Klein *et al.* 1968) and even in certain 'threshold' measures like critical flicker-fusion frequency (Thompson 1967).

These laboratory demonstrations that mental efficiency varies in a continuous manner right round the clock have been confirmed by observations of performance levels on different shifts in actual working situations, notably by Browne (1949) with switchboard operators, and Bjerner *et al.* (1955) with gas-works log-keepers. However, cumulative records of efficiency at different times during a particular shift are not readily obtainable in many industries, and gross overall figures may be misleading. Therefore the present author and his colleagues studied performance at certain tasks continuously during different 'shifts' in the laboratory over periods of 12 consecutive days (Colquhoun *et al.* 1968 a and b, 1969). It was found that, in general, mental efficiency followed the course of the body-temperature rhythm whether or not the latter was altering as a result of adjustment to the new sleep-waking routine imposed by working at unusual hours. On 'night' shifts this alteration consisted primarily of a 'flattening' of the rhythm, particularly during that part of it coinciding with the working spell. The flattening was accompanied by a parallel disappearance of the on-shift decrement in performance evident in the first part of the trial period.

Since the rate at which the temperature rhythm can be 'flattened' is slow it follows that the optimum theoretical solution for 24-hour working would include a 'fixed' night shift, the members of which would live permanently on an inverted sleep-waking routine. However, because such an arrangement would in all probability not be acceptable for social and other reasons, it is necessary to continue, by experiment, to seek for ways of investigating the probable detrimental effects on efficiency of the rapidly rotating systems which are likely to be adopted by most shift-workers in the future.

References

AARONS, L., 1968, Diurnal variations of muscle action potentials and word associations related to psychological orientation. *Psychophysiology*, **5**, 77–90.

BJERNER, B., HOLM, A., and SWENSSON, A., 1955, Diurnal variation in mental performance. *British Journal of Industrial Medicine*, **12**, 103–110.

BLAKE, M. J. F., 1967 a, Relationship between circadian rhythm of body temperature and introversionextraversion. *Nature, Lond.*, **215**, 896–897.

BLAKE, M. J. F., 1967 b, Time of day effects on performance in a range of tasks. *Psychonomic Science*, **9**, 349–350.

BROWNE, R. C., 1949, The day and night performance of teleprinter switchboard operators. *Occupational Psychology*, **23**, 1–6.

CHILES, W. D., ALLUISI, E. A., and ADAMS, O. S., 1968, Work schedules and performance during confinement. *Human Factors*, **10**, 143–196.

COLQUHOUN, W. P., BLAKE, M. J. F., and EDWARDS, R. S., 1968 a, Experimental studies of shift work I: A comparison of 'rotating' and 'stabilized' 4-hour shift systems. *Ergonomics*, **11**, 437–453.

COLQUHOUN, W. P., BLAKE, M. J. F., and EDWARDS, R. S., 1968 b, Experimental studies of shift work II: Stabilized 8-hour shift systems. *Ergonomics*, **11**, 527–546.

COLQUHOUN, W. P., BLAKE, M. J. F., and EDWARDS, R. S., 1969, Experimental studies of shift work III: Stabilized 12-hour shift systems. *Ergonomics*, **12**, 865–882.

DRUCKER, E. H., CANNON, L. D., and WARE, J. R., 1969, The effects of sleep deprivation on performance over a 48-hour period. *Human Resources Research Office (U.S.A.)*, Report No. 69–8.

FIORICA, V., HIGGINS, E. A., LAMPETER, P. F., LATEGOLA, M. T., and DAVIS, A. W., 1968, Physiological responses of men during sleep deprivation. *Journal of Applied Physiology*, **24**, 167–176.

KLEIN, K. E., WEGMANN, H. M., and BRUNER, H., 1968, Circadian rhythm in indices of human performance, physical fitness and stress resistance. *Aerospace Medicine*, **39**, 512–518.

KLEITMAN, N., 1963, *Sleep and Wakefulness* (Chicago: UNIVERSITY OF CHICAGO PRESS).

KLEITMAN, N., and JACKSON, D. P., 1950, Body temperature and performance under different routines. *Journal of Applied Physiology*, **3**, 309–328.

LOVELAND, N. T., and WILLIAMS, H. L., 1963, Adding, sleep loss, and body temperature. *Perceptual and Motor Skills*, **16**, 923–929.

PFAFF, D., 1968, Effects of temperature and time of day on time judgements. *Journal of Experimental Psychology*, **76**, 419–422.

TAYLOR, P. J., 1967, Shift and day work: A comparison of sickness absence, lateness, and other absence behaviour at an oil refinery from 1962 to 1965. *British Journal of Industrial Medicine*, **24**, 93–102.

THOMPSON, C., 1967, An investigation of the daily activity in the normal electroencephalogram, and its relation to physiological and psychological circadian rhythms. *Unpublished M.Sc. thesis, University of Aston, England*.

THOR, D. H., 1962, Diurnal variability in time estimation. *Perceptual and Motor Skills*, **15**, 451–454.

WEDDERBURN, A. A. I., 1967, Social factors in satisfaction with swiftly rotating shifts. *Occupational Psychology*, **41**, 85–107.

WILKINSON, R. T., 1965, Sleep deprivation. In *The Physiology of Human Survival* (Edited by EDHOLM and BACHARACH) (New York: ACADEMIC PRESS), pp. 399–430.

WILKINSON, R. T., EDWARDS, R. S., and HAINES, E., 1966, Performance following a night of reduced sleep. *Psychonomic Science*, **5**, 471–472.

Health and Productivity

By D. D. Bryson

Imperial Chemical Industries Limited, Nobel Division, Ardeer Factory, Stevenston, Ayrshire, Scotland

A group of 96 women, 83 under 20 and 13 over 25, were studied to ascertain if some health factors related to their industrial productivity.

A method of determining productivity was postulated.

Daily output related closely with the annual output of the individuals studied.

The wide variation in individual daily output (10–34 batches) occurred in spite of employee selection.

Natural aptitude, length of service and experience in the process were thought to be the major factors affecting daily output.

Health factors could not be demonstrated as influencing daily output.

Anaemia and sickness absence did influence productivity by reducing the number of days worked, particularly as groups with the highest productivity ratings lose more time than the lower productivity groups. The reason for this increase in absence was thought to be complex and not entirely due to the increased effort demanded for a higher output.

The need to investigate the reasons for sickness absence was restated.

Use of the Factory Medical Department appeared to reduce the incidence of sickness absence.

1. Introduction

The rate at which British industrial productivity has grown has not been as great as that of several other industrial nations. Industrial growth has therefore not been sufficiently rapid to finance the rise in the standard of living which the populace has come to expect.

Britain's relative industrial decline is certainly not a new phenomenon, but it is true to say that there is a new awareness of our present position.

It is uncertain whether the relative drop in productivity is due to shortcomings in the industrial worker, or to mismanagement of the overall industrial unit. This lack of understanding has led to periods of mutual recrimination between the parties involved. However, a more constructive attitude of self-criticism appears to be emerging, with management realizing their need to employ new managerial techniques and skills and, similarly, there are also indications that employees are also beginning to see that their new role in industry requires a greater degree of personal responsibility than was customary in the past.

The health of the working population in general has been shown to have a significant effect on national economic growth (Denison and Poullier 1967), but the effect of health on the individual's productivity does not seem to have been closely studied in this country.

As a comprehensive health service has existed in Britain for over 20 years, any widespread degree of ill health could reasonably be expected to be absent, or, if present, at least widely recognized. Yet it has only recently been shown that in a group of young workers in South West Scotland 37 per cent of all females and 19 per cent of all youths were found to be suffering from remedial iron deficiency anaemia (Bryson 1968).

The present study was designed to look at some factors influencing industrial productivity, especially those factors relating to health.

2. Method

A chemical factory is not the easiest plant from which to obtain measurable indices of individual performances. Fortunately a group of workers does exist on the site where individual output was accurately known for purposes of payment. This consisted of 96 females engaged in the manufacture of a detonator component. Males employed in the same department could not have their industrial performance compared with the females, as the men were engaged in a maintenance rather than a productive capacity. The statistics used in this study were obtained in 1969 and refer to a 12 months' period ending in October 1969.

The bulk of the women studied (83) were aged between 16 and 20, with a few (13) older married women.

The following factors were obtained for the group studied.

Physical characteristics—height, weight and haemoglobin levels

Family and social factors—size of family, health of parents and stability of family

Industrial factors—length of service

Health indicators—frequency of primary visits to the factory Medical Department for non-traumatic illnesses over a 12-month period. The number of days lost, attributed to sickness and the frequency of absences in the same 12-month period

The workers studied were classified into 4 groups by comparing their daily output. The output amounts were chosen arbitrarily in blocks of 5 batches which spanned the normal range of daily production:

	Total no. of girls	Age < 20	Age > 20
Group I produced 10–14 batches	7	6	1
,, II ,, 15–19 ,,	34	31	3
,, III ,, 20–24 ,,	34	33	1
,, IV ,, 25–30 ,,	21	13	8

3. Work Area and Job Requirement

The job carried out by the groups studied is sedentary and repetitive, requiring manual dexterity rather than physical effort. It consists of soldering wires to a 'fusehead'—a device very similar to a match head mounted on metal pins.

The work area is a large shed with benches running along its long axis. Heating, lighting and ventilation are satisfactory. The girls are dressed in overalls of the same colour, creating, by and large, a functional and rather institutional ambience.

Each girl is given a pre-employment aptitude test and only those individuals who are considered to have the technical ability required are given this particular job. Care is taken to tailor the bench lay-out to each girl and left-handed girls are given the appropriate facilities.

Training is of the 'Sitting with Nellie' variety and, although only fully trained operators were considered in this study, it is interesting to note that 46 trainees at the time of the study had a daily output averaging 13 batches per day. Although the range of performance of the trainees was wide, 5–23 batches per day, the average output equalled that of Group I.

The process, like so many manufacturing functions, could be considered as being potentially boring and presenting few opportunities for personal satisfaction, and much of the process is now being mechanized. None the less, labour turnover in the younger groups is not very high, being less than 20 per cent per annum.

In the usual course of events girls move to other departments where the wages are higher, at the age of 18, but some who are unhappy at the prospect of entering the explosives area of the Detonator Department are allowed to remain with their younger colleagues. Women who leave to get married on occasions return to their old department on re-employment.

4. Results

4.1. *Factors affecting Productivity in a Group of Female Workers*

All employees at this factory are medically examined before being offered employment and examined thereafter at annual intervals until the age of 19. The initial examination is mainly directed at discovering active disease rather than measuring absolute or relative physical fitness.

The failure rate at this test is low, being less than 5 per cent, and applicants who have some abnormality are restricted in their employment rather than being refused work out of hand. The most frequent causes for medical restrictions are monocular vision, dermatitis and pulmonary complaints which may be aggravated by the atmosphere of a chemical factory.

Thus the group studied had previously been screened by the medical and education departments and were thought to be suitable for the task they were to undertake.

4.2. *Relationship between Daily Output and Productivity*

Difficulty can be experienced in trying to assess the relative ' productivity ' of workers. However, in quantitative or ' piece work ' tasks the daily output can normally be obtained without too much trouble at various periods throughout the year; as production in any one day may not give a true indication of their annual output, nor does it allow for a gradual change in output as skill improves.

Table 1 compares average daily output with ' productivity ', which was obtained by multiplying the average daily output by the number of days worked in a 12-month period.

In all but two cases the groupings by daily output coincided with the annual productivity. Two workers from Group III, one from each age group, failed to come up to the annual productivity level of Group II although their daily output warranted their inclusion in Group III.

Table 1. Comparison of average daily output with annual productivity of each work group

Production	Number in group	Average daily output in ' batches '	Productivity in ' batches '
Group I	7	13·9	2730
Group II	34	17·7	4010
Group III	34	22·3	4840
Group IV	21	29·0	6320

4.3. *Relationship of Anaemia to Productivity*

As many of the girls were known to have iron deficiency anaemia, testing for this is now a standard feature of our medical examinations.

The performance of girls who had been treated, or were being treated, for anaemia was compared with that of girls not affected by this complaint (Figure 1). Anaemia was taken to be a haemoglobin reading of under 12 G per cent on an Eel Haemoglobinometer.

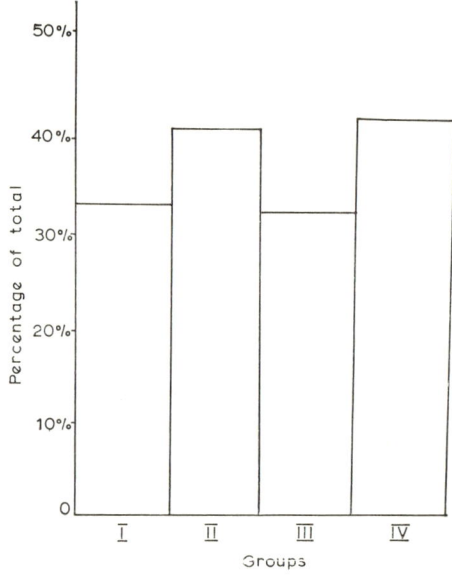

Figure 1. Proportion of girls in each production group with anaemia past or present

4.4. *Relationship of Illness and Productivity*

The question of the accurate assessment of the degree of illness experienced by an individual and the effect of this illness is known to be difficult and the measurement of fitness, or the lack of fitness can be even more difficult in the industrial environment.

This study used those parameters which were available as possible measures of the state of the individual's health. An obvious indication of illness would seem to be the amount of time lost from work, attributed to sickness over the year of study. Accepting the constraints that Taylor (1968) and others have demonstrated about sickness absence statistics and health, this measurement was still taken for those studied over a period of a year.

The frequency of such absences over the same 12-month period was also noted. Both these factors were related to the productivity of the individuals. Most, but not all, of the absences were certified by a family doctor. The absences recorded relate to total time off, i.e. including Saturdays and Sundays when illness lasts more than one week. The results are shown in Figures 2 and 3.

Also considered as a possible indicator of health was the frequency of first attendances at the Factory Medical Department for sickness (but not for accidents) over the course of the same 12-month period. The results of this review are shown in Table 2.

Table 2. Comparison of frequency of primary visits to medical department with (gross) absence from work attributed to sickness

Number of visits in one year	0–5	6–11	12–18	18–26
Gross absence attributed to illness	22.4	18.8	17.5	9.5
Number in group	54	28	8	6

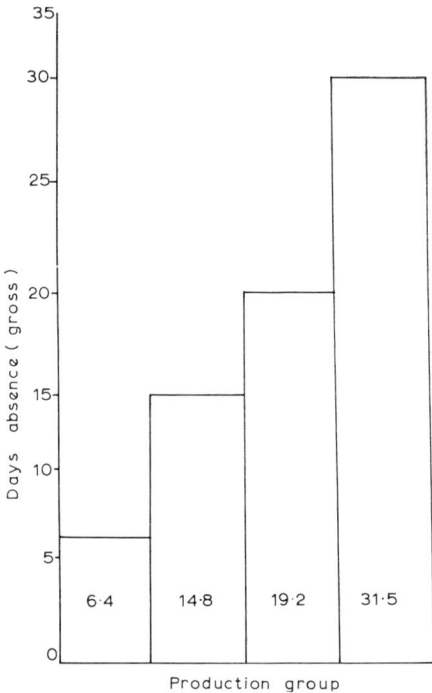

Figure 2. Mean number of calendar days lost in each production group

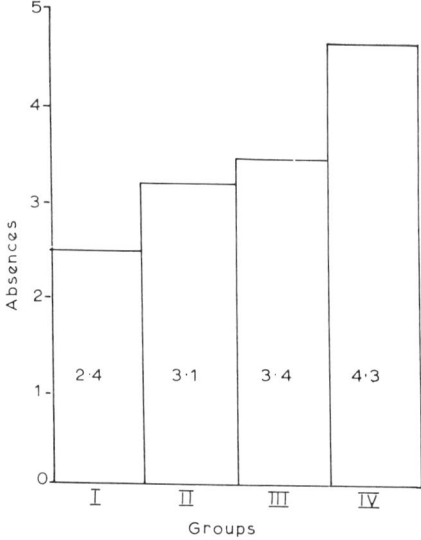

Figure 3. Mean spells of absence in a 12 month period for each production group

4.5. Non-Health Factors Related to Productivity

The length of service of each individual relates almost directly with experience at the particular job studied, as girls seldom return to this unit once they leave for other (more highly paid) departments, except as previously explained in the case of married women. Also there tends to be a direct link between age, productivity and length of service, as shown in Figure 4.

Comparison of performance between the younger and older groups could only be carried out in Groups II and IV, as I and II had only a single adult.

Nothing of importance emerged from the investigation to relate physical characteristics of the groups or the family size (number of siblings) in the younger group and productivity.

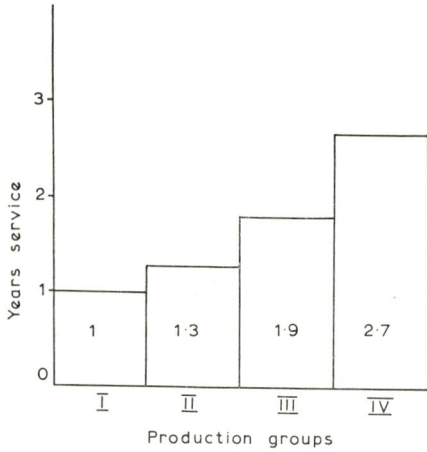

Figure 4. Average length of service in each production group

5. Discussion

If the quality factor is ignored, productivity could for the group studied be expressed by the formula

Productivity = Daily output × Number of days worked

Any factor which influences either the daily output or the number of days worked will obviously affect the productivity of the individual.

A very close relationship was found to exist between the workers' daily output and the annual productivity. Workers remained in the same productive groupings, except in the case of two individuals in category III, who, due to very long absences, would have required reclassification. Thus the productivity and the daily output give similar quantitative indications of the studied workers' industrial effectiveness.

Anaemia, past or present, did not appear to have any pronounced effect on the daily output, as the incidence remained fairly constant in all groups. It must be appreciated that the physical activity required of the girls was low, and a job where the physical effort was greater may well have produced more dramatic results. Also, these girls commenced treatment as soon as the anaemia was detected and repeated consultations with the Medical Officer may even have had some beneficial effect on their work (Mayo 1949).

Anaemia, however, does affect productivity in this factory by its influence on the number of days the affected person stays off work. As earlier work (Bryson 1968) demonstrated, the anaemic girls remained off work a greater number of days than girls with normal haemoglobin levels.

The average age of the 13 married women was 35 years and it was found that their absences tended to be longer in duration than the younger age group. There is a slight difference in the frequency of absence between the two groups, but this is not so pronounced.

Whereas the older women often have a higher than average daily output, their overall productivity does not compare so favourably with the younger group, who have similar daily outputs, because of the higher level of absence in the older (married) women.

It could be considered that illness which demonstrated itself in absence from work would be reflected in the frequent use of the Medical Department by those individuals most frequently absent. This correlation could not be demonstrated. Paradoxically, the highest number of consultations over a 12-month period was 26, and this girl, in fact, lost no time off work on account of sickness.

It would seem that girls who use the Medical Department frequently, as demonstrated in Table 2, have lower absence rates than the girls who do not consult the Medical Department so often. Those girls who by-pass the factory Medical Service either fail to seek any medical advice, or seek advice from their family doctor, and have the highest absence rates. Unfortunately, from the industrial point of view, they would appear to be in the majority in the group studied.

As the factory medical staff err on the side of caution when consulted, and since there is no hesitation in sending a patient home or to her family doctor, it is difficult to give a rational explanation for this behaviour pattern. Due to the geography of the area and a number of other factors, a close working relationship has developed between the factory medical staff and local practitioners. The role of both groups is seen to be similar in maintaining the health of the individual, be it worker or patient, yet their use by the patient appears to be different.

A possible cause of the failure by the workers to use effectively medical services provided by industry may well be due to a failure on the part of management and medical staff to publicize the services available to the workers and to explain the objects of the Medical Department.

Therefore where induction courses exist for new employees it would be valuable for members of the factory medical staff to be allowed to participate.

Individual productivity is most noticeably affected by a reduction in the number of days worked. The main stated cause of absence is sickness, whether certified or uncertified, and it is essential that the causes of absence be studied if productivity is to be improved.

Previous work (*Health of Munitions Workers*, 1917, 1918, Taylor and Pocock 1969) has shown many factors other than sickness, in the purely pathological sense, influence the length of time workers remain off work. In this factory several factors have been shown (Bryson 1968) to influence absence and these range from marital state, distance travelled to work, influence of the family doctor and the eligibility of the worker for sickness benefit.

A surprising feature of this study was the wide variation in the daily output of trained girls from a low of 11 to a high of 34 batches per day. This study suggests that absence seems to increase in step with increased output, the most productive workers staying off work for the greatest period.

The wide range in daily output seems to suggest that a natural aptitude, plus experience, is required to become expert in the job sequence studied. Natural aptitude must be considered as an important facet in productivity as some of the trainees could produce over 20 batches per day, i.e. more than the trained operators in Group I.

Perhaps the training methods could be improved and a continued selective weeding of girls, who find the operation excessively difficult, in spite of favourable aptitude tests, and replacing them in other jobs, would benefit both the worker and the Company.

I would like to thank Dr. P. J. Taylor for his encouragement and assistance in preparing this paper. Also Dr. J. D. Paterson, Division Medical Officer, Nobel Division, I.C.I., for permission to carry out the work leading to this publication. Finally, I would like to thank the personnel of the Ardeer Medical and Detonator Departments for their assistance in carrying out the work involved.

On a recherché sur un groupe de 96 femmes, dont 83 avaient moins de 20 ans et 13 plus de 25 ans, une éventuelle relation entre l'état de santé et la productivité.

Une méthode de détermination de la productivité est proposée. La production annuelle est étroitement liée à la production annuelle des individus étudiés.

La variabilité interindividuelle de production journalière est importante (10–34 lots) en dépit de la sélection des travailleurs.

L'aptitude innée, l'ancienneté et l'expérience de la tâche sont considérées comme facteurs d'influence majeur déterminant la production journalière.

L'existence d'une liaison quelconque entre critères de santé et production journalière n'a pas pû être mise en évidence.

Les congés pour anémie et maladie influencent la productivité en réduisant le nombre de jours de travail, en particulier du fait que les groupes ayant la productivité la plus élevée pendant davantage de temps que les groupes ayant la productivité la plus faible. Les motifs de cette augmentation de l'absentéisme sont complexes puisqu'elle ne semble pas uniquement liée à l'effort supplémentaire requis par l'augmentation de production.

La nécessité de connaître les motifs d'absentéisme pour raison de santé se fait à nouveau sentir.

La fréquentation du Service Médical d'Entreprise semble réduire l'absentéisme pour raison de santé.

Eine Gruppe von 96 Frauen, 83 unter 20 und 13 über 25 Jahre, wurde untersucht, um festzustellen, ob einige Gesundheitsfaktoren zu ihrer industriellen Produktivität in Beziehung stehen. Eine Methode zur Messung der Produktivität wurde aufgestellt. Der tägliche Ausstoss stand in engen Beziehung zur jährlichen Produktion der untersuchten Frauen. Die grosse Streuung des täglichen Ausstosses (10 bis 34 Einheiten) ergab sich trotz Einstellungs-Auswahl. Persönliche Geschicklichkeit, Länge der Dienstzeit und Erfahrung im Arbeitsprozess waren vermutlich die Hauptfaktoren, welche den täglichen Ausstoss beeinflussten. Gesundheitsfaktoren hatten dagegen keinen Einfluss auf den täglichen Ausstoss. Anämie und Fehlzeiten beeinflussten die Produktivität durch die Reduktion der Zahl der Arbeitstage, besonders weil Gruppen mit der höchsten Produktivität mehr Zeit verlieren als Gruppen mit niedrigerer Produktivität. Der Grund für diese Zunahme der Fehlzeiten war wahrscheinlich komplex und nicht allein dadurch bedingt, dass ein höherer Ausstoss eine grössere Anstrengung verlangte. Die Notwendigkeit, die Gründe für die Fehlzeiten zu untersuchen, wurde erneut betont. Die Mitwirkung des Werksarztes schien die Fehlzeiten herabzusetzen.

References

BRYSON, D. D., 1968, The incidence and effects of anaemia in a young industrial population. *Scottish Medical Journal*, **13**, 43.

BRYSON, D. D., 1968, Factors influencing sickness absence in young workers. *Transactions of the Society for Occupational Medicine*, **18**, 101–104.

DENISON, E. F., and POULLIER, J. P., 1967, Why growth rates differ: post-war experience in nine western countries.

HEALTH OF MUNITIONS WORKS COMMITTEE, 1917–1918, *CMND* 8511 *and* 9065 (London: H.M.S.O.)

MAYO, S., 1945, *The Social Problems of an Industrial Civilization* (London: ROUTLEDGE & KEGAN PAUL).

TAYLOR, P. J., 1968, Personal factors associated with sickness absence. *British Journal of Industrial Medicine*, **25**, 106.

TAYLOR, P. J., and POCOCK, S. J., 1969, Post-war trends in sickness absence and unemployment in Great Britain. *Lancet*, 1120–1123.

The Stresses and Strains on Feet in Industry

By R. J. Turvey

H. J. Heinz Co. Ltd., London N.W.10, England

In any group of people, 60 per cent have a foot anomaly; many only trivial, but some more troublesome. An endeavour is made to explain the detrimental effects upon feet of certain conditions associated with industrial activities. The manner in which foot disorders can be initiated or aggravated: by long sessions of standing heat and humidity special footwear is discussed. Mention is made of the added problems created in feet which exhibit structural deformities, and a recommendation is made regarding selection at pre-employment medical examinations. The value to the employees and to the industry of an industrial chiropody service is summarized.

1. Introduction

Statistics indicate that approximately 60 per cent of any group of people have a foot disability of some description. This may be so trivial as to cause no discomfort or anxiety at all, or it may be so gross that it will cripple the individual. Information to this effect was revealed by a survey made in 1966–67 by the Medical Care Research Unit of the Institute of Community Studies (Clarke 1969). Of those with trouble:

30 per cent had corns,
25 per cent metatarsalgia,
20 per cent nail trouble,
25 per cent other foot problems.

Many of these people are engaged in industrial activities, and I intend in this paper to give special consideration to them. Initially, therefore, I shall consider whether or not activity in industry aggravates conditions already present, or if certain conditions of employment create foot troubles. To this end I propose to deal with the effects on working people of the following conditions.

Lengthy periods of time standing or walking;
atmospheric and/or temperature variations;
the use of special footwear.

A number of other detrimental conditions can exist, but these three are sufficient to justify the proposal that foot disorders can arise from, or are aggravated by, working conditions.

2. Foot Disorders

2.1. *Standing or Walking*

Many workers spend a great deal of time standing by a machine, moving short distances to and fro, or walking long distances. Everyone will appreciate how any of these circumstances can give rise to tired, aching and swollen feet by the end of a day. Many people know only too well the symptoms which show themselves after they have spent a pleasurable day with the children at the Zoological gardens. But how much more complicated does the syndrome of symptoms become when the factory lady is middle-aged and overweight, or the employee is weight-lifting or load-shifting. It may be well to consider briefly why it is that feet begin to object when subjected to these ordeals. Does nature fail us? Are the natural constructional features of our feet so inefficient, or are the stabilizing and functioning influences on these features faultily designed? Are our feet unable to withstand the extra work-load? The fault rests squarely with us. Our environmental surroundings are to blame. Nature endowed most of us with a pair of well sprung, reasonably padded and highly adaptable pedestals and/or locomotory devices. We place these feet of ours in leather, wooden, rubber or plastic coverings, and leave them in these constricting circumstances for many hours each day. This is the custom and I am not endeavouring to advocate a return to nature, although we seem to be going that way in certain respects! It must nevertheless be realized that a great deal of intricate foot muscular control is lost by the majority of the adult population. When a worker is on his feet at his job all day, the extra strain can cause him to be additionally tense, and subconsciously more aware of his own discomfort than the demands of the job. This must create a lowered standard of efficiency. The distraction of thought increases the vulnerability to mistake and accident. Sometimes the foot is essential to the performing of work, as when it is required to operate a pedal or a treadle, or in work where it is necessary to bend and lift objects, where we retain our balance when the heel is raised, by the forward and outward spread of our toes. For this set of functions our toes need their suppleness of joints and the efficiency of their small controlling intrinsic muscles. If it is necessary for us to stand for long periods of time, the tissues within the soles of the feet are gradually compressed. This eventually reduces the shock-absorbing propensity of the natural fibro-fatty padding beneath the heels and the metatarsal heads. The loss of this cushion literally brings the bones nearer to the ground.

2.2. *The Effects of Atmosphere and Temperature Variations*

The body is very quickly affected by changes in external temperatures and humidity. These conditions may vary from hot to cold, from dry to moist; with combinations either way. The heat will activate the sweat glands, the humidity will slow the process of evaporation, and the feet soon show signs of hyperidrosis. This state arises from an engorgement of the peripheral blood vessels. The blood within those vessels endeavours to throw off the excessive heat by use of sweat glands. Prolonged overactivity of the sweat glands creates a vicious circle of events and the tiny muscles which control the orifices of the glands are liable to become so relaxed that the sweat literally runs out; in effect, the tap will not turn off. This will occur even when the individual is out and away from the hot moist atmosphere. From this it is plain that a state of affairs which arises from working conditions, unavoidable though they may be, can affect the individual at other times. The dampness of the sweat upon the skin is of itself very uncomfortable, but when the fatty acids within the perspiration are acted upon by the bacteria upon the surface of the skin, the objectionable and odorous condition of bromidrosis exists. This is an embarrassing state for the patient, and a very unpleasant one for everyone with whom he comes into contact. Very unpleasant though this state may be, a further complication can occur. The warm damp condition is just right to macerate the interdigital clefts and invasion by fungi completes the picture. The patient now experiences intense itching, and is, of course, a carrier of active fungi. This condition of dermatophytosis is very easily passed to other individuals, so here is a factor to be taken into account at pit-head baths and other places providing facilities for communal baths and showers.

Cold can create unpleasant symptoms, but unless the cold is really intense these are not of such difficulty to the patient. Conditions of cold experienced by workers exposed to the elements can give rise to foot symptoms, but it is easier for them to take precautions to guard against these. When the tissues are subjected to pressure and cold (a combination very liable on the soles of the feet) a blood stasis temporarily results. The affected part becomes oedematous, cyanotic and very painful, especially when subjected to the weight of the body. The plantar aspects in the heel region and beneath the metatarsal heads are especially susceptible to these effects. This is a state of chilling, in which there is a coagulation of the areas of affected tissue. In a healthy person the underlying tissues resolve the state and the area does not ulcerate. The resolution is a very slow process if the circulation is otherwise impaired. Cold may also have a drying effect upon the skin. This can create fissuring, especially around the perimeter of the heels. A dry skin on the sole of the foot can also form very tiny corns. These are called seed corns, and can give discomfort which may be likened to walking on gravel. The essential care in these cold conditions is to keep very dry coverings on the feet. The trans-Antarctic explorers, although combating extreme conditions, found at times, when sock-changing and boot-drying was impossible, that when eventually they were able to remove their foot coverings the superficial layers of the epidermis were 'shed like a cast' (Orr 1963).

2.3. *Special Footwear*

The occlusive properties of *rubber boots* brings this type of industrial footwear to the top of the list for the creation of foot symptoms. In addition to causing a hyperidrosis of the feet, they may cause the wearer to walk awkwardly. This is because the boots are weighty and large. Waterproof boots are now available in lighter-weight plastic materials, but the problem of excessive perspiration still seems to exist. Draining and drying these boots between shifts is very necessary. Using special racks, with a flow of warmed air, is most helpful. *Heavy, sturdy boots* are necessary to protect the feet of workers in foundries, other heavy industries, or for those who work on rough, uneven surfaces. This footwear certainly protects from external trauma, but may itself exert a traumatic effect upon the feet. Ankle spats and reinforced metatarsal guards are advised for foundry workers and welders. These guards save gross damage, but exert slight trauma upon the small bones which make up the mid-tarsal region of the feet and cause osteo-arthritic changes to occur gradually. After a number of years these bony changes give rise to pain or discomfort when walking. The tendons in the dorsum of the feet, to activate movement in the toes, are easily damaged. This is because this aspect of the foot is not endowed with very much fat. Although the plantar aspect of the foot is endowed with a more substantial fatty layer, jobs which require the use of the foot on a spade, the rung of a ladder, or even a foot pedal, can give rise to painful symptoms. The forces exerted on a particular part of the foot will cause a periostitis to arise on the pressurized bones. A band of tough fibrous tissue is present in this part of the foot, and among other functions it affords protection to the intricate system of tendons, blood vessels, nerves, etc. This so-called plantar fascia can cause crippling pain if seriously inflamed. It is therefore extremely important to take great care and give this part of the foot as much protection as possible. It is with this thought in mind that we move on to the somewhat difficult question of *old shoes for work*. How easy it is to be critical of the worker who uses his old and worn shoes for work. It seems logical to so many people, especially with the rising cost of footwear, to use shoes which have become too worn or shabby for ordinary wear. This is, of course, a very bad and dangerous practice. Some workers are astounded when it is suggested they should buy special footwear to use in the factory. When the already dilapidated shoe has been worn for a few weeks at work it becomes saturated with oils, the shank of the shoe is probably completely broken, the heels very badly worn down, and holes may be present in the soles. The looseness of the shoe provides little or no support for the foot, the worn heel places a strain on the ankle ligaments, and the sole allows sharp swarf or other dangerous substances to cut through to the foot easily. Many employers offer special *industrial footwear* and provide attractive payment schemes which encourage the employees to purchase it. As much publicity as possible should be given to this aspect of foot care.

The foregoing should provide some explanations of the special stresses and strains upon the feet of the industrial worker. A few of the conditions which may arise have been mentioned. The following section will deal with a number of the specific foot disorders which may be encountered.

2.4. *Valgus Foot*

This condition is frequently referred to as ' flat foot '. It shows itself by an inrolling of the longitudinal metatarsal arch. It is usually caused by the failure of ligaments to bind bone efficiently to bone, and by muscles weakening their hold and control of the foot. This is a real stress and strain symptom! The ' spring ' of the arch loses its effectiveness and the foot begins to hurt. The pain in the foot makes for a general tiredness in the individual. This state may arise as the result of long periods of standing or sessions of overstrain from heavy manual employment. Treatment needs to be provided as soon as possible. The condition should not be allowed to become chronic. The chiropodist can advise concerning the right type of footwear. If the patient is co-operative in this respect, a little help with padding and strapping may bring about a return to normality. If the condition has been neglected it may be necessary to make a special shoe inlay. This can be made with leather and rubber or fibreglass. By means of this appliance the bones are re-aligned and the muscles regain their correct action.

2.5. *Splaying of the Forefoot*

This is another instance of stretched ligaments. The transverse ligament should hold the heads of the metatarsal bones at their correct degree of spacing. The alignment with the toes should be straight and weight should be taken evenly over the forefoot. Stretching of the ligament causes the forefoot to become an unstable unit and the body weight becomes unequally distributed. This state of affairs gives rise to pressure lessions beneath the overburdened metatarsal heads and where the first and fifth metatarsal heads exert a pressure against the side of the shoe. The chiropodial treatment aims to provide protection and a more even distribution of weight.

The condition of valgus foot often combines with splaying and makes for a disorder which can create very painful symptoms. Certain inherited or congenital anomalies of the feet often predispose to either or both of these conditions, and a careful assessment of the feet at pre-employment medical examinations should be made. The information would aid the placing of employees, thereby giving a greater assurance that the individual has feet potentially suitable for the job envisaged. Employees should also be encouraged to seek early advice if symptoms manifest themselves. A certain amount of foot health advice should be given through the medium of works magazines, posters and the factory first-aid or medical department.

2.6. Heel Conditions

The heel can be a source of great difficulty. Deep-seated pain may be caused by the formation of bony outgrowths from the calcaneum. This occurs at the point of the attachment of the plantar fascia. Any undue strain or trauma in this area may cause bone cells to be produced, and a spur of bone will grow forward into the fascia. In very intractable cases it is necessary for a surgeon to excise the piece of bone, but early treatment will prevent such a stage being reached. Shearing and compressional stresses may cause a certain amount of breakdown of the fibro-fatty tissues which cushion the heel bone. This in turn will give rise to pain from a periostitis of the bone. Cushionings of rubber or foam may be introduced into the heel of the shoe to counteract these conditions, but occasionally it is necessary for the worker to change his type of job or vary the method by which he does it.

2.7. Hallux Valgus

This is a deformity affecting the first metatarsal bone and the great toe. The space between the first and second metatarsal heads increases, and the transverse ligament in that section becomes stretched. The medial aspect of the head of the first metatarsal becomes very prominent and thereby subject to traumatic damage. The consequent angulation of the great toe towards the second toe causes a variety of symptoms. Many troublesome deformities of the toes are to be seen on such feet and are due to the tremendous pressure exerted by the angled great toe. It would seem the condition originates from an inherent weakness in the foot, but may be very much aggravated by footwear with inadequate toe-box accommodation. This is why the feet of middle-aged and elderly ladies are frequently affected by the condition. Apart from the abnormal points of pressure and weight-bearing created by the instability of this important unit of the foot, the protuberance of the metatarsal head can cause a bursa to form and inflame. The area is also liable to ulcerate. The condition is sometimes corrected surgically, but protective replaceable appliances prepared by the chiropodist may prevent the condition from causing pain.

2.8. *Toe Nails*

Conditions affecting the toe nails are frequently seen by chiropodists in industry. For the worker who has dropped an object on his great toe, and caused a haemorrhage to occur beneath his toe nail, prompt attention is a great blessing. The chiropodist can painlessly drill the nail, remove the blood, and the pain is almost instantaneously relieved. The toe nail is saved and the nail bed does not undergo disturbing changes. If denied prompt and effective attention the condition will give pain for a long period of time and the nail will eventually be shed. The replacement nail may grow with deformity. In places where there is the employment of teenage boys the condition of ingrowing toe nails is frequently seen. This is a condition where there is a definite splinter of nail penetrating into the nail groove. This actually causes an ulceration with the associated problems of hypergranulation and infection. Boys of the age mentioned seem very susceptible to the condition. This is due to (*a*) greater liability to trauma on the toes, (*b*) the likelihood of a hyperidrotic type of foot, and (*c*) poor attention to the correct cutting of nails. The chiropodist can usually remove a section of the nail, together with the splinter, and give speedy relief. Many slightly older people may experience a lot of discomfort from the thickened and involuted type of nail, causing undue pressure. The nails may become thickened by a number of causes, but this thickness may be easily reduced and the involuted sides trained to a better shape. This is done by packing with various substances or by the use of spring-wire nail braces.

2.9. *Foreign Bodies in the Feet*

Many types of foreign bodies are found embedded in the feet. These may be hairs, bristles, pieces of glass, metal or wooden splinters. Even though they may be so small as to be hardly visible, they can give rise to a great deal of pain. It is remarkable how even long soft hair can wind its way into a sweat gland and become coiled within the skin. Many such hairs, two or three inches in length, have been removed. In certain industries the risk of such items entering the skin is quite high, and workers should be given every encouragement to take special care. They should also, of course, be advised to seek aid if any discomfort is felt. Prompt removal avoids the likelihood of unhappy complications.

2.10. *Verrucae and Infections*

In any community where people may be using changing rooms, shower baths, etc., special attention must be paid to hygiene. Even though strict care may be taken, verrucae and various fungal infections may easily occur and spread. Treatment, initiated in the early stages, effects a quick cure for the individual and reduces the resulting number of contacts.

The conditions mentioned are in no way intended to be a comprehensive list of disorders occurring in the feet, but to convey a broad outline.

3. The Chiropodist in Industry

Information has already been given concerning some of the special conditions to which feet are subjected, and various disorders have been briefly described.

The advantage to be gained by chiropodial relief from symptoms is not likely to be denied by any person who has made use of the opportunity of such a service. The Ministry of Labour, after an inspection of the welfare facilities in fourteen British factories (Ministry of Labour, 1961), wrote of the employees' enthusiastic comments regarding the chiropody service provided in some of the premises visited. The employees quite definitely appreciated the extra welfare facility. That this treatment benefits the recipient is beyond denial, but that it is worthwhile to the employer still needs to be proven. It has already been mentioned that:

(a) Treatment is on the premises. The employee does not take time to travel a distance. The time away from the work area is therefore reduced to the minimum.

(b) The individual's mind is quickly relieved of the diverting influence of the condition. Thoughts are much more likely to be focused on the job.

(c) Certain foot injuries may be promptly treated and complications prevented.

(d) Assessments may be made before people are placed on jobs for which their feet are not really suitable.

All four factors create benefits for both employer and employee. The former expending a small amount of money in return for a worker whose mind should be able to concentrate and who is able to spend more time doing the job for which he is paid. The latter, quickly relieved of painful symptoms, and endowed with a feeling of well-being, is appreciative of a good medical service.

Dans n'importe quel groupe de la population, 60 pour cent des individus présentent une anomalie des pieds; la plupart sont banales, mais certaines entraînent des troubles. Une tentative est faite pour expliquer les effets défavorables sur les pieds de certaines situations liées au travail industriel. Les auteurs discutent des différentes causes susceptibles d'engendrer de tels troubles: station debout prolongée, chaleur et humidité, port de certaines bottes. Mention est également faite du problème supplémentaire créé par l'existence de malformations congénitales, qui rend nécessaire une certaine sélection lors de l'examen méducal d'embauche. Les arguments montrant l'intérêt, à la fois pour le personnel et l'entreprise, de l'existence d'un service de chiropodie propre à l'entreprise, sont brièvement énumérés.

In jeder Gruppe von 60 Personen haben 60% einen anomalen Fuss: manche nur wenig, aber einige störender. Es wird versucht, die schädigenden Wirkungen gewisser industrieller Tätigkeiten auf die Füsse zu erklären. Die Art wird besprochen, wie durch langfortgesetztes Stehen, durch Hitze und Feuchtigkeit und durch besondere Fussbekleidung Fussschäden entstehen oder verstärkt werden können. Auch das zusätzliche Problem, das durch strukturelle Deformationen entsteht, wird erwähnt. Medizinische Fuss–UNtersuchungen vor der Einstellung werden empfohlen. Für die Angestellten und für die Industrie wäre ein chiropodischer Dienst sehr wertvoll.

References

CLARKE, M., 1969, Occasional paper No. 29, Medical Care Research Unit, Institute Community Studies (London: G. BELL & SONS LTD.).
ORR, N. W. M., 1963, Personal communication.
MINISTRY OF LABOUR, 1961, *Medical Services in Fourteen British Factories* (London: H.M.S.O.).

Session 2

Panel Discussion

Dummer, Beechams. Has Mr. Turvey any suggestion as to how a proper chiropody service can be provided for a small factory of, say, 1600 employees?

Turvey. A chiropodist can provide a service both for chronic and for acute disorders. An efficient chiropody service can be organized if a number of such factories combine to share the services of one practitioner.

J. C. Graham, Heinz. Our London factory has 2500 workers, and we employ a chiropodist on a sessional basis of two full days per month for day-time staff and one half-day per month for night-shift workers.

D. S. Ross, Babcocks. Can either of the first two speakers suggest to industrial doctors ways of creating useful records for research purposes, and ways to put the information available into a useful and meaningful form. How can one obtain the collaboration of experts in this field?

Colquhoun. One way would be to log the times of incidents. This might show a diurnal rhythm.

Bryson. Computerization could allow us to compile a complete record of a man's working life, including such things as exposure to toxic materials. Anyone considering formulating records for research purposes should gain the advice of a statistician at an early stage.

Malcolm. The Society of Occupational Medicine have a research panel which is at the service of anyone wishing to undertake work in this field: the panel can help in many ways, including knowledge of sources of information and experience in planning studies.

M. H. Davies, Fowler Davies & Co. Dr. Colquhoun drew attention to the effect of a psychological stimulus in nullifying the effects of diurnal rhythm changes. Has he any comment on the possibility of using physical stimuli to the same end?

Colquhoun. My colleague Mr. Blake tried using continuous noise in place of incentive, and obtained virtually similar results.

O. P. Llewellyn, Courtaulds. Is there any physiological reason for the observed temperature variations during the day? The apparent effects on performance may be connected with other physiological changes, such as falls in blood pressure, etc. I would also like to ask Dr. Colquhoun whether people on permanent night shift change their rhythm to give peak performance during their working hours?

Colquhoun. Our experiments indicate that if people could be placed on permanent night shift, their temperatures would even out. People with experience of night shift adapt more rapidly than others. With regard to the first part of the question, I did not mean to imply a causal relationship between temperature and performance. There are already 200 known physiological rhythms, and we as yet do not know which *cause* changes in performance.

D. P. Wyon, S.I.B., Sweden. I would like to ask Dr. Colquhoun just how mechanistic he feels the relationship between body temperature and performance may be. Even on the evidence he presented today one could draw the conclusion that there is no direct link. Does variation in body temperature from other causes affect performance?

Colquhoun. I agree with Dr. Wyon in saying that there is no mechanistic relationship between these two variables. With inspection tasks the detection efficiency varied with the body temperature: however, raising the body temperature above normal by physical work does change the performance, but the changes are not of the same kind as those which parallel circadian changes.

D. G. Davies, British Steel Corporation. Has Dr. Colquhoun any information regarding the effects of rapidly changing shift systems on indicators of body temperature and performance? Has he any knowledge about possible long-term effects of such changes?

Colquhoun. In brief, no. We have not looked at the 3.2.2 or the 2.2.2 systems. The nearest we have got to these is the naval system reported on in our paper. If such patterns are to be brought into use we should study them. One could argue from present knowledge that there is not likely to be any adaptation at all to such rapid changes, except perhaps in the very long term. However, adaptation to weekly systems is generally slow, and may be no better than with rapidly changing shift systems. With regard to long-term effects, this is not my field.

C. Lawrence-Jones, British Petroleum. Is there any indication of an optimal frequency of change in shifts?

Colquhoun. Purely from the limited point of view of working efficiency a permanent night shift is the answer. After 12 nights one's rhythm inverts but social reasons prevent continual night shifts for most people.

Session 3
Introduction

By J. R. Glover

The lack of research work on performance under sub-optimal conditions is reflected by the difficulty we had in finding speakers for this afternoon's session. It is well known that most epidemiological research that has been carried out in industrial medicine has been either on sickness absence certificates, e.g. lost time accidents regarded as being those where the man has lost three days or more from work: or, in the more serious diseases, the epidemiology has been carried out on death certificates, e.g. cancer of the bladder among aromatic amine workers, and asbestosis and mesothelioma among asbestos workers. It is extremely difficult to do quantitative work in industry on hazards that do not cause sickness absence, death, or acute physiological stress, and therefore we are fortunate this afternoon in having two speakers who have carried out work in areas where even the men themselves do not realize that they are working in sub-optimal conditions.

Dr. Geoffrey Smith is H.M. Inspector of Factories for South-West England and for some years he has carried out work on trichlorethylene and its effect on industrial employees. This afternoon he is giving us the details of his methods.

Professor P. J. Lawther is the director of the Medical Research Council's Air Pollution Research Unit at St. Bartholomew's Hospital Medical College. He is concerned with London smogs and their sulphur dioxide and sulphur trioxide content. He has chambers in which he can place human beings (volunteers) to study the effects of very low concentrations of sulphur dioxide. He is our leading expert in the medical profession in Great Britain on the subject of air pollution. It will be interesting to hear whether or not he has investigated the effect of very low levels of carbon monoxide in carbon monoxide chambers—if so, who he manages to persuade to be volunteers.

The Investigation of the Mental Effects of Trichlorethylene

By G. F. Smith

Department of Employment, 10 Gloucester Row, Clifton Down, Bristol BS8 4AT

1. Introduction

Nearly all investigations of the effects of trichlorethylene on the central nervous system have relied on the eliciting of symptoms from the subjects under investigation. The purpose of this paper is to describe personal experience with this method as well as to describe a second method of the investigation of possible mental effects designed to produce a result that avoids as far as possible observer bias; can be scored; and finally a result that can be compared with that from an unexposed control group. With symptoms, difficulties are encountered on each of these three points as will be explained later. Although during the day-to-day medical control of those exposed to industrial solvents such as trichlorethylene, the symptom must continue to be a valuable help; it does not lend itself readily to a research project, in which it will be required to demonstrate a measurable effect.

Trichlorethylene, being both a fat solvent and a narcotic, is thought to exert its major effect on the cerebrum. Usually this property is divided into the acute and chronic phases, although in practice it is difficult sometimes to separate the two, or indeed to be able to demonstrate that trichlorethylene has such a chronic effect. For the present it seems better to assume that a long-term effect due to some pathological or functional disturbance exists. Thus, the examination of those exposed at work to trichlorethylene may produce a picture which is the super-imposition of these two effects. In addition to this, what is more certain is that the metabolism of this solvent is so slow that such a situation will in any case occur. Thus one may be observing acute narcotic effects due to the trichlorethylene absorbed immediately before examination as well as that due to a narcotic metabolite resulting from the breakdown of trichlorethylene breathed-in perhaps two days earlier.

The prolonged metabolism of trichlorethylene has been demonstrated by several workers': by Powell (1945) on anaesthesia patients, by Gilchrest and Goldsmith (1956) on anaesthesia patients, and by Soucek and Vlachova (1960) on experimental subjects, and by Ahlmark and Forssman (1951) and Rivoire *et al*. (1962) on trichlorethylene workers. In all cases the slow trichloracetic acid excretion in the urine, after single doses and after continuous exposures, shown to extend over periods of approximately 21 days, was demonstrated. Those observing the continuously exposed such as degreasers and other workers with trichlorethylene reported a cumulative effect of the solvent once absorbed into the organism under such circumstances. Whether this slow elimination is due to the slow metabolic breakdown of the trichlorethylene from depôts of accumulation within the body, or whether it is due to the slow elimination of trichloracetic acid in any case through the kidneys, still remains undecided. Ahlmark and Forssman (1951 a) found that trichloracetic acid given by mouth to human subjects followed a similar excretion course as that trichloracetic acid formed by metabolic breakdown, namely a steep rise to peak values followed by a slow fall to low values. Paykoc and Powell (1945) noted that only after 10 days did 75 per cent of an injected dose of trichloracetic acid appear in the urine. Similar prolonged excretion rates have been found in animals (Barrett et al. 1936, Fabre and Truhant 1952).

Trichlorethylene, as well as being itself narcotic, forms narcotic metabolites. The foremost of these is trichlorethanol, which has a narcotic power similar to that of tribromoethanol (Avertin), although it is possibly more toxic. It is certainly more narcotic than trichloracetic acid which appears to have little toxicity at all (Hewer 1938, Mikiskova and Mikiska 1966, Case 1943, Butler 1948, Hewer and Hadfield 1941, and Ahlmark and Forssman 1951 a). Trichlorethanol is formed in greater quantities in the urine after the absorption of trichlorethylene than is trichloracetic acid, namely approximately 50 per cent of the absorbed trichlorethylene is excreted as trichlorethanol as compared with 20 per cent of trichloracetic acid, as found in human experimental subjects by such investigators as Bartonicek 1962, Soucek and Vlachova (1962) and Teisinger (1960). Mikishova and Mikiska (1960) demonstrated that a given amount of trichlorethanol raised the threshold of excitability of the motor cortex to electrical stimuli in guinea-pigs as much as 200 to 400 per cent of unexposed levels, whereas such a dose of trichlorethylene had no effect. Many workers have demonstrated that chloral hydrate is an intermediate product of trichlorethylene metabolism. Byington (1965) was one able to demonstrate this, although Marshall and Owens (1954) failed to do so, because their chemical method of detection was probably not sensitive enough to detect this substance, which is likely to have a very short biological half-life. Chloroform has also been claimed as a metabolite and has been found in human urine after trichlorethylene exposure, as stated by Lob (1960) and Mikiskova and Mikiska (1966). There seems little doubt therefore that the peculiar metabolism of trichlorethylene ensures both a prolonged and complex pattern of effect on cerebral tissue.

Grandjean and his colleagues (1955) were among of the few investigators who, as well as recording symptoms among trichlorethylene workers, submitted 50 of their chronically exposed subjects to mental tests. Reliance solely on the recording of the frequency of complaints was made by Frant and Westendorp (1950), and Ahlmark and Forssman (1949), to mention a few. Most authors undertaking such investigations did not compare their results with those from observations made on matching control groups. Grandjean and his colleagues (1955) and Bardodej and Vyskocil (1956) relied on the comparison with known normal values for the mental tests and on the statistical significance of the incidence of complaints considered typical of the neurasthenic syndrome, thought to occur with trichlorethylene exposure. Ahlmark and Forssman (1951 b) relied on comparison with the overall frequency of such complaints, as apparently recorded by official government bodies in the Scandinavian countries in which they carried out their observations on industrial workers.

2. Methods of Investigation

In all, 230 men and women exposed to trichlorethylene have been examined. In this group there are 108 men receiving tests of mental function, which will be described later in this paper. The 108 men were compared with 63 male controls working at the same factories, who matched the exposed groups for many different factors, such as age, marital status, etc. Although the factories in which they worked were not selected in a strictly random fashion, some having been chosen for investigation because high exposures were known to be taking place or at least were suspected; this applied only to a small proportion. The subjects concerned were asked for their reactions to the fumes of trichlorethylene, leading questions being avoided as far as possible. No attempt was made to distinguish between slight and marked symptoms, as had been attempted both by Ahlmark and Forssman (1951) and by Grandjean and his colleagues (1955). A similar technique was adopted for 112 men and women exposed to lead in low concentrations, as verified by haemoglobin levels of 14 G per cent and above and by blood levels of below 30 μg per cent for the men and by haemoglobin levels of 13·5 G per cent and above for the women. This group matched that exposed to trichlorethylene in age and sex distribution, and it was hoped that it would serve as a reference group for the determination of the likely frequency of the complaints commonly met in those exposed to trichlorethylene in workers neither exposed to any solvent nor to a high level of toxic substance in the course of their work.

3. Results

The symptoms encountered among the 130 subjects exposed to trichlorethylene were as in Table 1. Among the gastro-intestinal effects were nausea and occasional vomiting, flatulence and changes in bowel habit thought by the subject to have some or a definite connection with exposure. The effects described as 'autonomic' or as 'vegetative' by many authors were flushing of the face, sweating of the hands and palpitations. Unavoidably, the decision whether a symptom represented changes in the autonomic nervous system or

Table 1. Complaints—all subjects exposed to Trichlorethylene

Complaint	No	%
No complaint	74	27·0
Fatigue	97	74·5
Dizziness	73	56·2
Headache	23	17·7
Gastro-intestinal	33	25·4
Autonomic effects	10	7·7
Other	35	24·9

in any other symptom tended to be arbitrary. Grandjean et al. (1955) noted fatigue both at night and at work among 44 per cent of the workers he selected for investigation, and they found that 50 per cent of their subjects suffered from vertigo. Bister-Miel (1964) found that 36 per cent of the workers that she examined suffered from dizziness. Of the above 130 subjects, emotional changes were rare, but investigators such as Grandjean et al. (1955), Bardodej and Vyskocil (1956), Andersson (1957) and Ahlmark and Forssman (1951 b) found such effects as depression, inattention, perseveration of ideas, confusion and feelings of anxiety to a greater extent among the workers they examined. Alcohol and tobacco intolerance are usually considered to be characteristic of chronic trichlorethylene intoxication. Among the 130 subjects, who were not teetotal, 8·9 had intolerance to alcohol and the same proportion were intolerant to tobacco. This was evidenced by flushing of the face, a feeling of warmth in the body and tremors when small quantities of either were taken; similar findings were noted by the authors mentioned above.

For the purposes of this investigation, trichloracetic acid in urine level was accepted as a fair index of exposure to the solvent vapour. Of the 130 subjects seen, 60·8 per cent had levels up to 20 mg per litre of urine and, if 60 mg per litre were chosen, this percentage rose to 82·1 per cent.

The actual number of separate complaints was chosen as one way to assess the effect of exposure. It was found that those subjects with levels of up to 60 μg per litre had an average number of complaints of 1·8 per person, whilst above this level the average rose to 2·7. Up to 20 mg per litre this average was only 1·3, the difference between this average and that for the group having urinary levels of higher than 60 mg per litre was thought to be statistically significant. The average number of complaints in the group not exposed to solvents and who were found by laboratory investigation to be probably in good health had an average number of complaints of the same type of 1·3, therefore corresponding very closely to the low exposure group. Reference to individual results showed that there was considerable variation in any of the groups selected, this having been found particularly by investigators such as Frant and Westendorp (1950), many in the highest exposure group having no complaints, whilst many in the lowest group had 6 or even more. In many such cases it was known that industrial circumstances other than exposure were likely to have produced anomolies of this nature.

Next, the type of complaint was matched against the low (less than 60 mg per litre of trichloracetic acid in the urine) and the high exposure group (more than 60 mg per litre of trichloracetic acid in the urine). The following was found as described in Table 2.

Table 2. Type of complaint against trichloracetic acid level in the urine

Complaint	Less than 60 mg/l		More than 60 mg/l	
	No	%	No	%
No complaint	69	31·2	5	12·2
Tiredness	78	41·2	24	58·5
Dizziness	58	30·6	18	43·9
Headaches	18	9·5	2	4·2
Gastro-intestinal	17	9·0	7	17·1
Autonomic effects	4	2·2	1	2·1
Other	31	16·3	12	29·2

It was evident during the course of the investigation that some account would have to be taken of whether a complaint was slight or marked, as had been found to be necessary by other investigators, notably by Ahlmark and Forssman (1951 b). The results obtained by doing this are shown in Table 3. It will be seen that there was close correspondence in the proportion of subjects in either group admitting to only slight symptoms. Again, the responses may not all have been determined entirely by exposure.

Because such workers as Granjean et al. (1955) thought that the total length of exposure to trichlorethylene had a significant effect on the incidence of complaints, the average number of complaints per person in each of four total exposure length groups was worked out as shown in Table 4. In the case of the subjects under discussion, the total time of continuous exposure to trichlorethylene appeared to have little influence on the frequency of symptoms.

Table 3. Severity of complaint against trichloracetic acid level in the urine—'slight' complaint

Urinary trichloracetic acid excretion mg/l	No	%
Up to 60	48	25·3
More than 60	9	22·0

Table 4. Number of complaints against total length of exposure

Total length of exposure	Average number of complaints	Average trichloracetic acid level in the urine mg/l	Number of subjects in the group
Less than 1 year	1·3	35·1	51
Up to 5 years	1·5	26·2	112
Up to 10 years	1·8	20·9	47
More than 10 yrs	1·4	35·8	37

4. Discussion

Reliance upon symptoms as a means of measuring effect on the nervous system involves dependance upon subjective impressions not only on the part of the examiner but also on the part of the subjects of the investigation. The results presented indicate that symptoms probably result from exposure, at least at higher levels, but there is also the strong possibility that the symptoms observed result also from other factors in the working environment. This seems to point to a need for an objective means of assessing the effect of solvents such as trichlorethylene on the central nervous system. The same need was felt by both Grandjean et al. (1955) and by Bardedej and Vyskocil (1956). Weitbrecht (1965) went as far as suggesting that tests of mental function should be included in the medical examination of workers exposed to the solvent. As with symptoms, however, it would seem that the true function of such test is to validate an index of exposure such as the trichloracetic acid level in urine on a sufficiently large number of subjects, rather than to apply them to individual cases. On cases of acute exposures, Chalupa and his colleagues (1960) carried out memory testing on cases of trichlorethylene gassing.

The following tests were carried out on 108 of the exposed subjects and on a group of 63 unexposed controls:

Cornell Medical Index Questionnaire, Psychiatric Section.
Heron's Personality Questionnaire.
Fluency test.
Concentration and Attention Test—13-Mistake Test.
Concentration and Attention—serial sevens and digital span.
Memory Testing—logical sequence, 2 tests and one delayed repeat; mechanical memory of word list, recognition memory and memory of words having associated meaning.
General Knowledge Test.

All these tests are capable of being scored, and many, such as the Cornell, and the questionnaires demanded ' yes ' and ' no ' responses, these having been filled in without the examiner being present. Care was taken with the other tests as well to avoid any help being given by the examiner or any influence that he might have. Many of the tests such as the digital span, serial sevens and logical memory tests were standard psychiatric testing procedures, but the 13-Mistake Test was introduced by Grandjean from Bleuler, and it was claimed that this exercise based on mental arithmetic was particularly likely to detect the mild psycho-organic syndrome that they claimed occurred as a result of chronic trichlorethylene exposure. The mechanical memory tests were used by Chalupa and his colleagues, and reduction of memory function was found among cases of gassing for a considerable period afterwards; normal values for these tests were obtained by these workers from large numbers of unexposed and healthy subjects. Although numerous, these tests were simple to do, and in no case took longer than 45 minutes to carry out. The plan of examination of both exposed and control subjects was to investigate their medical state, reactions to solvent and to enquire into their past occupational, medical and social histories before the lunch break, during which the questionnaires were filled in. After the lunch break, mental testing was carried out; usually in one day no more than three subjects were so examined. The exposed subjects were examined no earlier than Wednesday in the week unless they had been working through the weekend.

The Cornell Medical Index, the Personality Questionnaire and the General Knowledge test probably served more as a means of matching the control and exposed groups, the distribution of response in the two groups having turned out to be similar. This was an expected result.

5. Conclusions

The peculiar metabolism of trichlorethylene is described and its relationship to the effect on the central nervous system of the solvent outlined. The results from observing symptoms complained of among 230 subjects are given, and the likely value of assessing symptoms for detailed observation of the effects of trichlorethylene of the central nervous system is discussed, the main objection being the introduction of errors of subjective judgment that they introduce into a planned investigation. To avoid this, a system of objective mental testing is described as part of a research project into the effects of trichlorethylene at work.

I wish to acknowledge help given to me by Dr. T. A. Lloyd Davies, H.M. Senior Medical Inspector of Factories.

References

AHLMARK, A., and FORSSMAN, S., 1949, Theoretical background of the quantitative pyridine–alkali test for trichlorethylene intoxication. *Proceedings of the IXth International Congress for Occupational Medicine*, 552.

AHLMARK, A., and FORSSMAN, S. 1951 a, The effect of trichlorethylene on the organism. *Acta physiolica Scandinavica*, **22**, 326.

AHLMARK, A., and FORSSMAN, S., 1951 b, Evaluating trichlorethylene exposures by urinanalyses for trichloracetic acid. *Arch. industr. Hyg. Toxicol.*, **3**, 386.

ANDERSSON, A., 1957, Gesundheitliche gefahren in der Industrie bei Expositionen für Trichlorethylen. *Acta med. Scand. Supplementum*.

BARRETT, H. M., CUNNINGHAM, J. G., and JOHNSTON, J. H., 1936, A study of the fate in the organism of some chlorinated hydrocarbons. *J. industr. Hyg. Toxicol.*, **21**, 479.

BARTONICEK, V., 1962, Metabolism and excretion of trichlorethylene after inhalation by human subjects. *British Journal of Industrial Medicine*, **19**, 134.

BISTER-MIEL, F., 1964, Application des methodes modernes toxicologiques au problème de l'hygiène industrielle concernant la nocivité du trichloréthylène pour l'homme. *Thesis, University of Paris.*

BUTLER, T. C., 1948, Metabolic transformation of chloral hydrate. *J. Pharmacol. exper. Therap.*, **92**, 49.

BYINGTON, K. H., and LEIBMAN, K. C., 1965, Metabolism of trichlorethylene in liver microsomes. *Mol. Pharmacol.*, **1**, 247.

CASE, E. H., 1943, The present status of trichlorethanol. *Anesthesiol.*, **4**, 523.

CHALUPA, B., SYNKOVA, J., and SACEVIK, M., 1960, The assessment of electroencephalographic changes and memory disturbances in acute intoxications with industrial poisons. *British Journal of Industrial Medicine*, **17**, 238.

FABRE, R., and TRUHAUT, R., 1952, Contribution des études experimentales chez l'animal—II. *British Journal of Industrial Medicine*, **19**, 39.

FRANT, R., and WESTENDORP, J., 1950, Medical control of exposure of industrial workers to trichlorethylene. *British Journal of Industrial Medicine*, **12**, 131.

GILCHREST, E., and GOLDSMITH, M. W., 1956, Some observations on the metabolism of trichlorethylene. *Anaesthesia*, **11**, 28.

GRANDJEAN, E., MÜNCHINGER, R., TURRIAN, V., HAAS, P. A., KNOEPFEL, H. K., and ROSENMUND, H., 1955, Investigation into the effects of exposure to trichlorethylene in mechanical engineering. *British Journal of Industrial Medicine*, **12**, 131.

LOB, M., 1960, L'action du trichloréthylène sur le taux d'alcool dans le sang. *Medicina Lavoro*, **51**, 587.

MARSHALL, E. K., and OWENS, A. H., 1954, Excretion and metabolic fate of chloral hydrate and trichlorethanol. *Bull. John Hop. Hosp.*, **95**, 1.

MIKISKOVA, H., and MIKISKA, A., 1966, Trichlorethanol in trichorethylene poisoning. *British Journal of Industrial Medicine*, **23**, 116.

PAYKOC, Z. V., and POWELL, J. F., 1945, The excretion of sodium trichloracetate. *J. Pharmacol. exper. Therap.*, **92**, 49.

POWELL, J. F., 1945, Trichlorethylene: absorption, elimination and metabolism. *British Journal of Industrial Medicine*, **2**, 142.

RIVOIRE, J., GÉNÉVOIS, M., and TOLOT, F., 1962, L'elimination de l'acide trichloracétique chez les sujets exposes au trichlorethylene. *Arch. Mal. prof.*, **23**, 395.

SOUCEK, B., and VLACHOVA, D., 1960, Excretion of trichlorethylene metabolites in human urine. *British Journal of Industrial Medicine*, **17**, 60.

TEISINGER, J., 1960, The maximum allowable concentration of trichlorethylene in air. *Proceedings of the 13th International Congress for Occupational Health.* New York, 987.

WEITBRECHT, U., 1956, TRI und TRI-Ersätze in der Metallindustrie. *Zeut. f. Arbeitsmed. u. Arbeitsschutz*, **15**, 138.

Carbon Monoxide and Phenobarbitone: A Comparison of Effects on Auditory Flutter Fusion Threshold and Critical Flicker Fusion Threshold

By A. D. L. Guest, Catherine Duncan and P. J. Lawther

Department of Pharmacology and MRC Air Pollution Unit,
St. Bartholomew's Hospital Medical College, London

In a small series of experiments, eight healthy subjects breathed either air alone or a mixture of air and carbon monoxide designed to raise carboxyhaemoglobin saturations to approximately 10 per cent. Neither subjects nor operators knew which gas was administered, and the effects of the exposures on auditory flutter fusion thresholds and critical flicker fusion thresholds was assessed during a period of 6 hours after the end of the exposure. There was no evidence of any depressant effect of carbon monoxide on the thresholds: the auditory flutter fusion threshold increased a little with carbon monoxide.

To check the response of the subjects to a drug that had already been tested, the experiments were repeated, with oral administration of phenobarbitone and a placebo. For the group as a whole there was a depressant effect of phenobarbitone on both of the thresholds, as expected.

1. Introduction

Much is known about the symptoms and clinical effects associated with exposures to carbon monoxide producing concentrations of carboxyhaemoglobin in the blood equivalent to saturations of 20% COHb and over. Carbon monoxide is commonly present as a pollutant of the air in streets and in workplaces in concentrations which may lead to saturations considerably less than those producing symptoms; there have been some attempts to investigate possible effects of these blood levels on perception or the performance of fine tasks. In the investigation reported here subjects were gassed to produce saturations of approximately 10% COHb and effects sought by measuring auditory flutter fusion thresholds (AFFT) and visual critical flicker fusion thresholds (CFFT), both of which have been shown in man to be sensitive indicators of depression by centrally acting drugs (Besser, Duncan and Quilliam 1966, Besser 1966, 1967, Simonson and Brozek 1952). In both tests the interrupted noise and light are perceived as fused when the frequency of interruption is greater than a critical value, the fusion threshold. The trials were double blind cross-over with balanced treatment orders. Using air as the placebo gas statistical analysis of the results demonstrated that CO had no effect on the values tested, and in order to determine whether the subjects studied were 'normally susceptible' a further series of tests was done in which phenobarbitone and a placebo were administered; phenobarbitone (100 mg) has been shown to depress auditory flutter fusion threshold (AFFT) and visual critical flicker fusion threshold (CFFT) (Besser and Duncan 1967).

2. Methods

2.1. Gassing

The method of administering the gas was carefully chosen. Calculations were made to determine the volume of carbon monoxide needed to produce saturations of approximately 10% COHb. Obviously, had this amount of gas (about 150 ml) been exhibited in high concentration in air, 'boluses' of blood containing relatively large concentrations of COHb would be transported to the brain with each inspiration; on the other hand if the brain were never to receive concentrations in excess of 10% COHb the subjects would have to be exposed to weak mixtures for periods of time in excess of 8 hours at rest before equilibrium between blood and inhaled gas could be reached. In these experiments a compromise method was used whereby the desired saturations were acquired by breathing at rest for 65 minutes from a bag containing 250 ml CO in 500 litres air. Respired volume obviously varied with size of subject and this roughly compensated for variation in blood volume so that fairly uniform blood levels resulted from the inhalations.

The eight subjects were healthy male volunteers (checked by medical examination), aged between 23 and 48 (mean 41), and all but one were non-smokers. Experiments were always started at the same time of the day (09.30). CO and air, and later, drug and placebo, were administered at intervals of 7 days between treatments. Gas from a bag (air or 500 ppm CO in air) was administered by mouth-piece or face mask to the subjects lying on a couch. The volume of expirate was measured and concentrations of CO were monitored and determined by use of a non-dispersive infrared analyser. Percentage saturation of the blood samples with carboxyhaemoglobin was determined by the micro method of Commins and Lawther (1965) before and immediately after gassing and again at $1\frac{1}{2}$, 3, $4\frac{1}{2}$ and 6 hours after the end of the exposure. Respiratory rate, expired volume and pulse rate were measured at intervals throughout the gassing. The rate of dissociation of COHb in the blood at rest when breathing air is equivalent to a half-life of 4 hours (Hackney, Kaufman, Lasher and Lynn 1962), and in order to keep the elimination of CO as nearly standard as possible subjects were told to avoid exercise during the day of the test. All subjects abstained from coffee, tea, alcohol, drugs and cigarettes from 10 p.m. the previous evening until the end of the test.

In the second series of trials phenobarbitone 100 mg was given in a capsule, as was a matched placebo of identical appearance. The subjects were told the nature of the drug being studied, but elaborate precautions were taken to ensure that neither they nor the experimenters knew the allocation of active or placebo treatments. The codes were disclosed for statistical analysis only after completion of the series.

2.2. Auditory Flutter Fusion Threshold

A modification of the technique of Miller and Taylor (1968) was used for measuring thresholds.

The apparatus for producing AFFT consisted of a transistorized noise generator. This produced a white noise (i.e. containing frequencies from under 500 c/sec to over 6000 c/sec) which was interrupted at a 9 to 1 ratio, the rate of the interruption being determined by an oscillator which could be adjusted in steps of 1·0 interruptions/sec from 15 to 100 i/sec. The interruptions were switched on and off for periods of 1 sec so that the signal presented binaurally to the subject was composed of alternating 1 sec sequences of continuous and interrupted noise with no intervening gap.

The signal was transduced by a pair of earphones TDH 39 fitted with rubber ear muffs, the intensity of the noise as delivered by the earphones in these experiments was 57 dB (re. 0·0002 dyne/cm² as generated in a 6 cc rigid acoustic coupler). The response of the earphones was linear over the frequency range 500–6000 c/s.

All experiments were conducted in a sound-proof booth. The subject was presented binaurally with the stimulus and instructed to say ' yes ' if he could clearly distinguish the repeated bursts of interrupted noise. If he could not, or if there was any doubt, he was to say ' no '. The starting interruption rate was 100 i/sec, well above fusion threshold for all subjects. The frequency interruption was reduced in steps of 10 i/sec until a point was reached 10 i/sec above that of the subject's threshold (as judged from a previous measurement on that subject). The frequency was further reduced in steps of 1 i/sec until the AFFT was passed. The AFFT was taken to be the fastest interruption rate at which the auditory flutter (i.e. interruption) could be clearly distinguished and below which the ' interrupted ' responses were given consistently. The descending sequence of interruption rates was used in all experiments. The subjects were not told the technique used, but all had been introduced to the tests in practice runs, prior to the main experiment.

2.3. *Critical Flicker Fusion Threshold*

A continuously flickering neon light was used driven by a rectangular pulse generator with a mark space ratio of 1 : 1 and which illuminated an area of 0·4 cm². Subjects were seated with their eyes on a level with the light source and 60 cm from it in a sound-proof booth.

The subject was presented with the light stimulus and instructed to say ' yes ' if he could see that the light was flickering. If he could not or if there was any doubt he was to say ' no '. The starting frequency was 53 i/sec, well above fusion and this was then reduced in steps of 10 i/sec until a point was reached 2 i/sec above the subject's previously measured threshold. Then the frequency was further reduced in steps of 0·25 i/sec until the CFFT was passed. The CFFT was taken to be the highest interruption rate at which the interrupted responses were given consistently. As in the case of the determination of the AFFT, the descending sequence of interruption rates was used in all experiments. Each set of observations consisted of 3 AFFT and 2 CFFT determinations, starting with measurements of AFFT and then alternating. The end point of the AFFT determination was the more difficult of the two for the subjects to recognize and therefore the first AFFT determination of the set was used as a practice run.

The means of the two remaining AFFT and two CFFT determinations in each set were used for analysis. A set of measurements was made immediately before treatment as a base-line. In the first series the subject then inhaled the CO mixture or air, and as soon as this was complete another set of measurements was taken, $1\frac{1}{2}$ hours after the base-line test, and again at 3, $4\frac{1}{2}$, 6 and $7\frac{1}{2}$ hours after it. In the second series the subject then took drug or placebo and readings were taken at $1\frac{1}{2}$, 3, $4\frac{1}{2}$, 6 and $7\frac{1}{2}$ hours later.

3. Results

Pre-exposure COHb% saturations varied between 0·3 and 2·5 per cent (mean 0·91). Immediate post-exposure values were between 7·8 and 11·2 per cent (mean 8·85, S.D. 1·08). Curves of decay were, with two minor irregularities, as expected, and in the samples taken 6 hours after the end of the exposure, COHb saturations varied between 2·9 and 4·5 per cent (mean 3·43, S.D. 0·52). There were no appreciable variations in pulse rate or respiration during exposure. Subjects were asked whether or not they had symptoms they attributed to the exposure. In the one case in which definite symptoms were reported the exposure preceding their development was to air. A few subjects felt drowsy after the drug/placebo exposure, but this symptom was unrelated to the nature of the preparation taken. Some subjects said that they found the tests difficult and some aggression was aroused in a few of the subjects. In one subject who was withdrawn from the experiment before the formal trial began the testing procedure provoked an attack of migraine.

The mean starting values on the two tests (Table 1) remained virtually the same throughout the series, but there were large differences between and within individuals, and the effects of the 'treatment' administered on any given day were assessed in terms of the difference from the starting value.

Table 1. Mean starting values for AFFT and CFFT

	AFFT (i/sec)	CFFT (i/sec)
Air	37·25	35·62
Carbon monoxide	34·56	35·64
Placebo	36·44	34·06
Phenobarbitone	36·88	35·28

Changes in threshold during the $7\frac{1}{2}$-hour test period are shown in Figures 1 and 2. In each of the two series of control experiments there was an increase in the threshold for the AFFT and a slight decrease in that for the CFFT. These trends are similar to those reported in other series of experiments with drugs (Besser and Duncan 1967), and, as in the earlier work, the effect of the 'active' drug has been determined from the differences between the change in threshold with the active and placebo treatment. Results were worked out individually, and the mean 'effects' of phenobarbitone and carbon monoxide on the eight subjects are shown in Table 2.

Table 2. Mean effect of carbon monoxide and phenobarbitone on AFFT and CFFT in eight subjects

Time (hours) after first test	Carbon monoxide				Phenobarbitone			
	AFFT		CFFT		AFFT		CFFT	
	Mean	S.E.	Mean	S.E.	Mean	S.E.	Mean	S.E.
$1\frac{1}{2}$	+3·06	2·26	+0·246	0·514	−3·94	2·65	−1·245*	0·645
3	+4·38†	1·82	+0·050	0·527	−7·19*	3·37	−2·146†	0·645
$4\frac{1}{2}$	+3·00*	1·58	−0·034	0·438	−5·63	3·22	−2·122†	0·799
6	+0·94	2·39	+0·278	0·673	−5·44	3·34	−1·378*	0·673
$7\frac{1}{2}$	+3·06	3·15	+0·852	0·817	−4·13	3·41	−1·459*	0·744

* $p \leqslant 0.10$ † $p \leqslant 0.05$

The effect of phenobarbitone was similar to that found in earlier series: the thresholds for both tests were depressed at each time interval after the administration of the drug and the depression induced in the AFFT was greater than

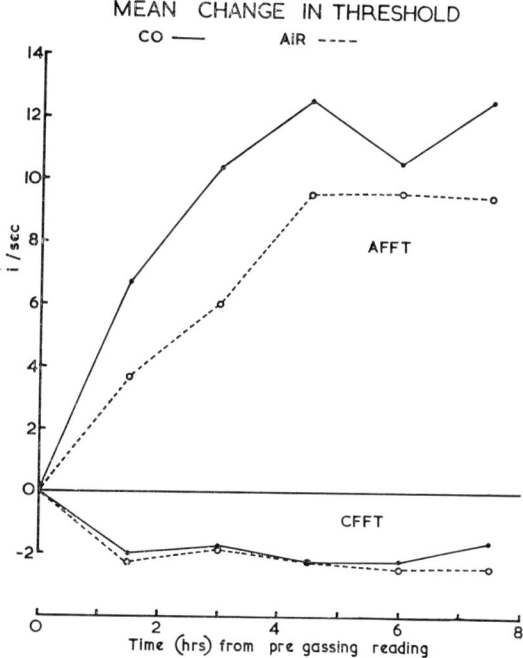

Figure 1. Mean change in AFFT and CFFT following carbon monoxide and air in eight subjects.

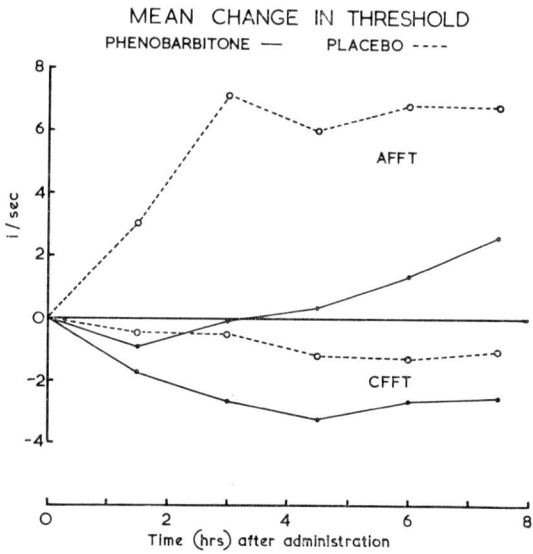

Figure 2. Mean change in AFFT and CFFT following phenobarbitone 100 mg and placebo in eight subjects.

that in CFFT. With only eight subjects the standard errors were however large, and the 'effect' of phenobarbitone in the group only reached the 5 per cent level of significance (Student's t-test) at two points, and for the two CFFT measurements made at 3 and $4\frac{1}{2}$ hours.

Carbon monoxide did not produce any depression in the AFFT or CFFT. As shown in Figure 3, there was a tendency for the AFFT threshold to increase rather than decrease, and there was no effect at all on the CFFT. The increase in AFFT reached the 5 per cent level of significance only at the 3-hour point.

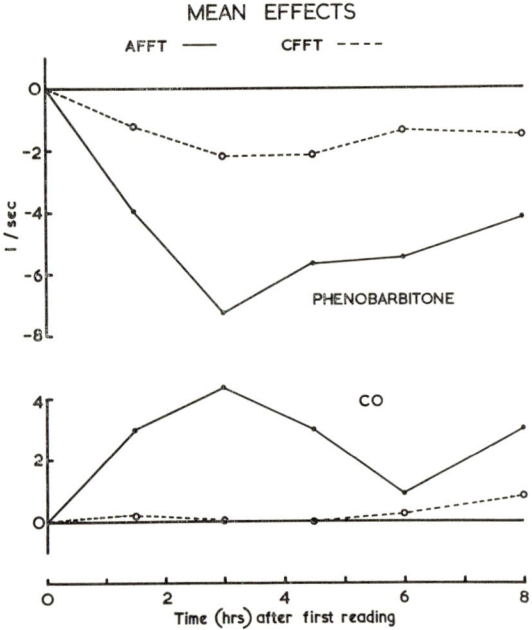

Figure 3. Mean effects of carbon monoxide and phenobarbitone on AFFT and CFFT in eight subjects.

In view of the large scatter within this small group, the results were examined further on an individual basis. Four of the eight subjects showed a marked effect of phenobarbitone, but the other four did not. Since the object of doing the phenobarbitone experiment in this series was to check whether the subjects responded in a normal manner to a standard drug, the possibility that only four of the subjects could be regarded as 'reactors' was considered. The results were re-analysed for the two sub-groups 'reactors', and 'non-reactors', but this did not affect the findings on carbon monoxide at all: both reactors and non-reactors showed some increase in AFFT and only very small changes in CFFT.

4. Discussion

The results of these experiments do not indicate any depressant effect of COHb saturations of approximately 10% on AFFT or CFFT thresholds. Since only eight subjects were studied, small changes could have escaped notice, but in the case of AFFT there was a consistent 'improvement' rather than a deterioration in threshold.

Von Post-Lingen (1962) showed that when the concentration of carboxy-haemoglobin was greater than 16 per cent a significant fall in threshold of CFFT occurred. In our experiment none of the levels of carboxyhaemoglobin was above 16 per cent. Double blind conditions were not used in Von Post-Lingen's study and it is possible that her procedures might have influenced some of the results obtained. Schulte (1963) used various tests such as the colour stimulus response test, letter stimulus response test and plural noun underlining test, and his analysis of the results indicated that impairment occurred in these tests with levels of carboxyhaemoglobin below 5 per cent. One particular criticism of this study is the unexplained finding that in 28 observations concentrations greater than the equilibrium of 14·0% COHb (Forbes et al. 1945) were recorded in the blood after gassing with 100 ppm of carbon monoxide. The tests were not double blind, and no mention is made of whether stimulants such as tea or coffee were permitted before or during the tests; such considerations might influence the results obtained with the tests used by the author.

Beard and Wertheim (1968) found that low doses of carbon monoxide (50 ppm inhaled at rest for 90 minutes) appeared to impair the ability of subjects to discriminate between different durations of sound. No values for the concentrations of carboxyhaemoglobin in the blood are given since the analyses of the blood samples collected for this determination were stated to be inaccurate. Measurements of carboxyhaemoglobin are of vital importance in assessing the exposure to carbon monoxide and the omission of such measurements in the paper seriously limits its value. These authors, as did Schulte (1963), did not specify whether double blind techniques were applied or whether stimulants were used.

Any effect of moderate concentrations of carbon monoxide in the blood is of interest in relation to drivers of motor vehicles in busy streets, although it has been shown that cigarette smoking rather than exposure to traffic fumes is the main factor in determining COHb saturations (Lawther and Commins 1970). Whilst the work reported here has not provided any evidence of an effect of COHb saturations of about 10 per cent on two tests that have been used successfully to demonstrate effects of cerebral depressant drugs, further work is in progress using other tests of perception and performance, and this will be reported in a later paper.

Our thanks are due to Professor J. P. Quilliam for his encouragement, to Mr. R. E. Waller and Dr. J. McK. Ellison for their valuable advice, to Dr. B. T. Commins and our technical officers for their analyses of blood and gas, and to Mr. J. D. Gasking who designed the electronic equipment used to determine the auditory flutter and critical flicker frequency thresholds. The authors also wish to express their appreciation to the Medical Research Council and to the Governors of St. Bartholomew's Hospital for research grants to Professor J. P. Quilliam for some aspects of this work, including support for ADLG (MRC), CD (St. B.H.) and J. D. Gasking (St. B.H.).

Au cours d'une courte série d'expériences, on a fait respirer à huit sujets en bonne santé, soit de l'air pur, soit un mélange d'air et d'oxyde de carbone afin d'entraîner une élévation d'environ 10 p. 100 du taux de carboxyhémoglobine. Ni les sujets, ni l'expérimentateur ne savaient quel gaz avait été administré. Une fois l'nhalation terminée, on a recherché, au cours des 6 heures consécutives quels pouvaient en être les effects sur les seuils de fusionnement du papillotement auditif et du papillotement visuel. Aucune détérioration n'a pu être mise en évidence sous l'effet de l'oxyde de carbone, si ce n'est une faible élévation du seuil de fusionnement du papillotement auditif.

Afin de vérifier les effets d'une drogue qui avair déjà été utilisée dans une expérimentation antérieure, ces expériences ont été répétées, mais en faisant ingérer aux sujets du phénobarbital ou un placebo. Comme prévu l'ingestion de la drogue a entraîné un effet dépresseur sur les deux types de seuils.

In einer kleinen Serie von Versuchen atmeten acht gesunde Personen entweder Luft allen oder eine Mischung von Luft und Kohlenmonoxyd, um die Carboxyd-Sättigung auf annähern 10% zu bringen. Weder die Versuchspersonen noch der versuchsleiter wussten, welches Gas geatmet wurde. Die Wirkungen der Beatmung auf die auditive Flatter-Verschmelzungs-Frequenzschwelle und die kritische Flimmer-Verschmelzungs-Frequenzschwelle wurden währens einer Periode von sechs Stunden nach Beendigung der Beatmung bestimmt. Es zeigte sich keine senkende Wirkung der Kohlenmonoxydatmung auf die Schwellenwert: die auditive Flatter-Fusions-schwelle nahm unter Kohlenmonoxyd ein wenig zu. Um diese Reaktion der Personen auf einge Droge, die schon getestet worden war, zu prüfen, wurden die Experimente mit oraler Verabreichung von Phenobarbiton und einer Placebo wiederholt. Für die Gruppe im ganzen ergab sich ein depressiver Effekt auf beide Schwellenwerte wie erwartet.

References

BEARD, R. R., and WERTHEIM, G. A., 1967, Behavioral impairment associated with small doses of carbon monoxide. *American Journal of Public Health*, **57,** 2012–2022.

BESSER, G. M., 1966, Some physiological and psycho-pharmacological studies on auditory flutter fusion and qualitative weight perception in normal subjects. Two new psycho-pharmacological tools. *M.D. Thesis, University of London.*

BESSER, G. M., 1967, The time course of action of diazepam. *Nature*, **214,** 17–19.

BESSER, G. M., and DUNCAN, C., 1967, The time course of action of single doses of diazepam, chlorpromazine and some barbiturates as measured by auditory flutter fusion and visual flicker fusion thresholds in man. *Brit. J. Pharmacology and Chemo*, **30,** 341–348.

BESSER, G. M., DUNCAN, C., and QUILLIAM, J. P., 1966, Modification of the auditory flutter fusion threshold by centrally acting drugs in man. *Nature*, **211,** 751.

FORBES, W. H., SARGENT, F., and ROUGHTON, F. J. W., 1945, The rate of carbon monoxide uptake by normal men. *American Journal of Physiology*, **143,** 594–608.

HACKNEY, J. D., KAUFMAN, G. A., LASHIER, H., and LYNN, K., 1962, Rebreathing estimate of carbon monoxide hemoglobin. *Archives of Environmental Health*, **5,** 300–307.

LAWTHER, P. J., and COMMINS, B. T., 1970, Cigarette smoking and exposure to CO. *New York Academy of Sciences* (in press).

MILLER, G. A., and TAYLOR, W. G., 1948, The perception of repeated bursts of noise. *Journal of the Acoustical Society of America*, **20,** 171–182.

SCHULTE, J. H., 1963, Effects of mild carbon monoxide intoxication. *Archives of Environmental Health*, **7,** 524–537.

SIMONSON, E., and BROZEK, J., 1952, Flicker fusion frequency. Background and applications. *Physiol Review*, **32,** 349–378.

VON POST-LINGEN, M-L., 1962, An experimental study of the effect of carbon monoxide on critical flicker fusion and Evipan tolerance tests in healthy persons. *Orebro, Sweden.*

Session 3
Panel Discussion

D. S. Ross, Babcock & Wilcox Ltd. Could I ask the last speaker how long after exposure were the various samples taken? We have a problem with carbon monoxide exposure of welders, but cannot take samples at once.

Lawther. The half-life at rest of carboxyhaemoglobin is four hours, so the initial concentration can be estimated from later tests. However, the results of such an extrapolation will be affected by exercise and any other atmospheric pollution, so it is best to take a sample as soon as a man comes off his shift if possible.

A. H. Jones, British Rail. With regard to Dr. Smith's comments on the significance of subjective complaints such as headache, we found that people unknowingly working in an atmosphere contaminated with trichlorethylene, and when using new paint stripper, complained of symptoms similar to those he described.

Smith. Similar symptoms occur, of course, with many different solvents.

Hockey, Rolls-Royce. Scottish workers using trichlorethylene have complained of an odd feeling in the scrotum.

D. S. Ross, Babcock & Wilcox Ltd. Workers exposed to low concentrations have complained of frequency of micturition which disappeared when exposure ceased.

Smith. I have not heard of scrotal effects elsewhere, though there have been complaints of impotence. Work here is being done on the 17 ketosteroid output in this context. Trichlorethylene may have drastic effects on the skin if it is in high concentration. Exceptional cases of renal damage have been reported, but these are rare.

W. H. A. Beverley, B.B.A. Group. Have those exposed to trichlorethylene been retested after a month or more? Also was there any difference in persistence of symptoms in non-alcoholic controls where compared with the others?

Smith. Retesting after an interval is difficult, firstly as many people may have left the factory, and secondly because of the possible effects of learning on the second series of tests. We looked at the alcohol situation and found no difference between drinkers and non-drinkers.

R. Murray, T.U.C. One early symptom of narcosis from trichlorethylene and white spirit in painters working in confined conditions is known as 'the music'; the workers hear a pop group or brass band and realize it is time to stop and get some fresh air.

Smith. I have not met this, but it could well occur in the pre-narcotic excitement stages.

D. P. Wyon, S.I.B., Sweden. What sort of effects on performance could one expect from a 10 per cent level of CO saturation in the body? Flicker fusion and flutter tests demonstrate altered levels of arousal, and the second series can only show high stress levels. These various tests are themselves arousing, and thus small subtle changes in performance may not be detected.

Lawther. Only one of the CO trials was significant at $p = 0.05$. 10 per cent saturation gives no effect on flutter or flicker-fusion. My adviser, Dr. Broadbent, considered that the second series of tests were most sensitive in relation to driving.

D. P. Wyon, S.I.B., Sweden. While I am aware of the difficulties with flicker-fusion, I wonder what alterations in performance are likely to be detected by Dr. Wilkinson's tests.

Wilkinson. My recordings of the behavioural effects of stress shows that most situations showing changes are unpleasant ones. Unpleasantness and boredom will effect test performance, and thus we tried to make our tests as interesting as possible. The vigilance and five choice tests are not stimulating after the first five minutes, and can be very boring after 45 minutes. Just because they are not stimulating they should reveal moderate effects of stress, and Professor Lawther's negative results are therefore of considerable interest, particularly since these tests have been shown to be sensitive in other situations.

Session 4
Introduction

By Dr. R. Murray

As a member of each of the three societies participating, I think it is particularly important that joint meetings of this type occur regularly. The shape of health is changing in medicine; initially concerned with the treatment of established disease, the physician, with the help of his biochemical and physiological colleagues, is concerned now with the detection of presymptomatic signs of disease. This move towards positive health can occur only if there is teamwork between the members of our different societies.

With reference to its title, the word performance makes me think of this conference as a symphony in four movements. We started with the woodwind of Dr. Atherley, and have proceeded through the various movements until we come now to the final movement. This is to some extent a reprise of the *motif* with which we started yesterday, with some emotional overtones.

Studies of Children under Imposed Noise and Heat Stress

By D. P. WYON

Statens Institut for Byggnadsforskning, LTH–A Sektion, Box 725, S–220 07, Lund 7, Sweden

> An assessment is made of moderate stress research as a field of study in itself, in which principles governing the choice of criteria of stress and strain are reviewed. The most important parameters of moderate stress research are identified and placed in the context of an empirical scheme, with reference to which their interpretation and relevance are discussed. Studies by the author of children under imposed heat stress are used to illustrate the scheme and are interpreted in terms of arousal and effort. An approach to the study of noise in the moderate stress region is suggested, in which measures based on sound pressure level are unlikely to have much relevance. This view is supported by preliminary results from a study of children under intermittent low-level noise.

1. Introduction: Moderate Stress Research

This conference was intended to focus on human reactions to minor changes in the environment, whatever these changes are and however the reactions happen to affect performance and behaviour. We are here to discuss 'moderate stress' as an area of study in itself. To achieve this, the organizers chose the admirably elastic title of 'performance under sub-optimal conditions', thus avoiding the need to define what is moderate, and by making it a joint meeting, avoided any preconceptions of what is stress. It is unfortunately the case that heat physiologists, for example, may meet and talk about heat stress, but may never attend the same conferences as acoustic engineers who meet to talk about noise stress. If they did, they would doubtless find themselves in a separate section.

This has been satisfactory up to a point, as interest has naturally been directed first towards the effects of extreme stress, which are in fact often stress-specific. High levels of noise cause pain and damage to the ears and interfere with speech communication. Heat stress does not do that. High levels of heat stress limit the amount of physical work possible and can overload the heat balance mechanisms to the point of collapse. Noise stress does not do that. Pressure, vibration, illumination, the ionization of the air—all have their specific effects which are demonstrable at extreme levels of stress. They are naturally the first objects of study, and form the topics of discussion in the respective sections of their respective specialist conferences. This has been satisfactory, but I believe that it is no longer so. It is a question of the criteria to be used for optimization.

Some of the stress-specific criteria evolved for extreme levels of stress can be characterized as unacceptable above or below a certain limit, and this at once defines the maximum tolerable range, but not all of them have even this property. Thus body temperature is stable within a certain range about the optimum, and unstable when the capacity of the regulative mechanisms is exhausted. Sweat rate is the most important of these mechanisms, and has a maximum that serves to define the upper limit of stable body temperature; it has also an 'elbow', where the sweat rate begins to rise more rapidly with temperature than before, but this cannot be considered to form the upper limit of some ' desirable range of temperature ' all in cases. Many people do not in the least mind sweating; their skin temperature is lower and they can drink more. In the special case of drinks being in short supply, sweating is perhaps undesirable in the long run.

The point to be made is that even such apparently useful delimiters as this are not independent of other conditions, and are often not in themselves systems-relevant. Just as onset of sweating is an irrelevant criterion for the idle drinker, so speech interference is an irrelevant criterion for the solitary worker. Criteria must be systems-relevant to be usable for the definition of any optimum or neutral zone at all. Quite frequently the so-called neutral zone is merely the range in which no variation in the chosen criterion can be shown. Such a zone can be very wide, as in the case of stable body temperature or physiological damage to the eyes and ears. The resort is then to subjective judgment, and much effort goes into derivation of the dose–response relationship. This is usually for the wrong reason, namely, that the subjective response of a group varies about the optimum within the otherwise neutral zone.

Subjective response is studied as a last resort and not because it is relevant or useful to define an optimal range. It is expressed as ' comfort ' because this sounds as if it is not stress-specific. Subjects therefore often have difficulty interpreting the questions. When the results come to be used, we have the familiar round-table vote as to how many people shall be permitted to be uncomfortable within the optimal range. By the time subjects have voted on what they think we mean by comfort, and planning committees have voted on how many they think should be comfortable, the process begins to look more like a democratic institution than a scientific technique. Like all democratic institutions, this one has its lobbies. The clients of this type of research have usually some very good economic reasons for extending the optimal range, and the final round-table vote is inevitably a reflection of this fact.

Moderate stress research needs criteria that are more sensitive than the ones that delimit the physically tolerable range, and needs criteria that are systems-relevant rather than stress-specific. I want to suggest some of the ways in which this can be achieved by means of a discussion that is applicable to moderate stress of all kinds, but which was derived from my own work on children under imposed heat and noise stress. Hence the rather specific title of my contribution. The clients of my research have been building engineers and architects, the subjects children, the criteria mental performance and behaviour. The aim has been to derive a range of sensitive and relevant delimiters for heat and noise conditions inside buildings.

2. Systems-Relevant Criteria of Stress and Strain

Research, and particularly applied environmental research, has tended to concentrate on deriving empirical relationships between selected parameters of stress and strain—or dose and response, the terminology depending on the professional allegiance of the researcher. It has been a long task, and it has really only just begun. Each new set of circumstances seems at present to need an empirical investigation, providing indefinite employment for researchers in the field ,but otherwise undesirable. To progress it is necessary to consider in depth the processes operative between receptors and effectors, and I shall emphasize below why this is particularly so for the moderate stress region.

Before considering the 'intervening variables' that mediate stress–strain relations, I would like to make some observations about the choice of parameters of stress and strain. The category of parameters of stress may be taken firstly to include strictly physical factors of the environment, directly observable, independent of the subject, and for the most part capable of very accurate quantification. That accurate quantification of the environment in terms of conventional physical parameters is not sufficient, and may therefore not be necessary, is one of the main conclusions I shall draw in the course of later discussion.

The category of parameters of strain may be taken to include all the directly observable responses of the subject to his environment-physiological measures, subjective reports, observed behaviour and performance of defined tasks. From these, the choice of criterion in any study must be governed by the objective of the system under consideration. At least three levels can usefully be distinguished in the hierarchy. They may be characterized, in ascending order, as internal, individual and total systems.

At the lowest level, interest may be centred on one or more of the various internal systems of the body. The system objective can be defined as continued satisfactory functioning, and the chosen parameters of strain must be relevant to this objective. Specialists in the various disciplines can usually deal independently with the stress–strain relationships relevant to this level of the hierarchy. However, these internal systems, even taken together, form only a sub-system of the individual as a whole.

When interest is centred on this next system level, that is, on the conscious interaction of an individual with his surroundings, the system objective is different, and no longer constant with time as it was for the internal sub-systems. Comfort and well-being are usually adequate system objectives at this level, but under some circumstances they may be secondary to other goals, as when discomfort and distress are readily endured in the course of duty or sport, or even for less heroic reasons. Quite wrong conclusions may be drawn by assuming too readily that system objectives are fixed at the individual level, especially if they are stress-specific. Thus sports cars are not chosen for their low noise level or efficient heating, though their owners may at other times lay great emphasis on these factors. Windows are still popular and often opened, despite their apparently detrimental effects as measured by most comfort criteria for heat, noise and lighting. At this level, then, system objectives must be much more carefully identified than for the internal sub-systems. It must be borne in mind that the simultaneous satisfaction of sub-system objectives is a necessary condition, even though it is not a sufficient condition at the individual level.

Strain measures based on system considerations for the individual will therefore usually delimit a much narrower range of acceptable stress than will strain measures based on internal sub-system considerations, and for this reason are often used in moderate stress research. Their disadvantage is that their systems-relevance is not constant with time.

To overcome this disadvantage, interest can be centred on the system objective of the next level of the hierarchy, namely, the efficient functioning of the total system, of which the individual forms only a part. The total system may be a school, office, factory or social group. System objectives are usually much better defined at this level, and are usually constant with time, often because they form the *raison d'être* of the unit. The most important strain measures relevant to this level are performance measures, and I place them firmly at this level. Whenever the individual is performing a task, he is functioning as a part of the total system, and there are many factors—like money—which can influence his objectives and which are only present in the context of the total system.

I find that careful distinction between these three system levels can be a very valuable antidote to confusion. Vague concepts such as stress, strain and indices thereof become concrete when related to the appropriate system. Take indices of thermal stress, for example. MacPherson (1962) lists nineteen in common use, and I must confess to having unwillingly added to the list (Wyon *et al.* 1968). Physical parameters such as temperature, vapour pressure and air movement are usually combined in some way and related to some measure of strain. The system under study might be the internal thermoregulatory system, and sweat rate would be a suitable index of strain in this sub-system. However, if we should wish to study the next system level, some subjective estimate of comfort, such as Mean Thermal Vote, might be chosen as an index of strain. Sweat rate could then be an entirely suitable index of stress in this system, as the widely-used P4SR index recognizes. If we should wish to study the total system, some performance measure might be chosen as the index of strain. Mean Thermal Vote could then in turn be a more suitable index of thermal stress than the predicted sweat rate. The original combination of physical parameters could at each level be used to predict the indices, but need not necessarily itself be the most appropriate index of stress nor need it always show the closest empirical relationship to the chosen index of strain. It is easy to apply the same reasoning to other sources of stress. One just has to remember that what is strain for the goose may be stress for the gander. In any multi-disciplinary subject, and especially in ergonomics, this simple principle can go a long way towards clearing up misunderstandings and differences of emphasis between specialists.

3. Intervening Variables

I have dealt at some length with what we should call factors we can confidently define and measure. I should like now to deal with how we should define and measure factors that we can confidently name, but which prove elusive when any attempt is made to use them for purposes other than verbal discussion. I should like to call them intervening variables, on the black-box analogy, in the sense that they are parameters of the intervening processes that occur to link changes in environmental stress with our chosen indices of strain.

These processes are intrinsically of academic interest, but we should not as ergonomists be much concerned with them if we could always obtain straightforward relationships between stress and strain, relationships that enabled simple regression lines to be drawn, large correlations satisfyingly approximating to unity to be calculated, and accurate predictions to be made.

It is the lack of such textbook simplicity in the field of moderate stress that is forcing ergonomists to consider intervening variables. With no clear idea of what the intervening processes are, not even a functional model of their operation can be put forward with any confidence, far less a mechanistic model.

Nevertheless, we do postulate a large number of intervening variables to account for the variance obtained in our studies, diagnosing their existence and nature by studying the empirically available evidence, rather than by an appeal to a comprehensive theoretical model. It is this derivation from evidence that makes it useful to distinguish the following three categories, working down the chain of causation from the parameters of environmental stress: time-variant factors closely linked to stress, time-invariant factors independent of stress and strain, time-variant factors closely linked to strain.

3.1. *Time-Variant Factors Closely Linked to Stress*

We know very little about how perceptual information is transmitted and processed, but it is safe to assume that the physical parameters derived from studies of heat, noise and light propagation outside the black box have very little descriptive power within it. We do not know the appropriate dimensions, so we cannot devise experiments to quantify the parameters. The ' little man ' at the receiving end does not know them either, but in psychophysical experiments he can return fairly meaningful answers relevant to the intensity dimension that is so appropriate for description of the world outside, and has so often been the chosen dimension for stress research. Using his answers, we can correct for the difference between objective and subjective intensity, as we do when weighting decibel measures, for instance, and as we do *not* yet do for temperature and many other stress factors. Objective level of intensity has proved a valuable parameter of stress at high levels, when the effects are clearcut but we shall almost certainly have to apply the appropriate psychophysical corrections at lower levels of all stresses. Using Mean Thermal Vote as a stress index in performance experiments, as suggested in an example above, would be one way to make this correction for heat stress.

The only other dimension we can begin to distinguish is information content. Here there is an even greater distinction between quantification outside and inside the black box. Objective probabilities can be handled by information theory, but subjective probabilities are more important and much more difficult to study. The concept of ' expectancy ', as applied to the well-defined events in vigilance experiments, has proved useful (e.g. Taylor 1967). In such experiments the subject's past experience and present interpretation of the situation can be controlled. They must certainly play a very large part in determining subjective probabilities in real-life moderate stress situations, such as dealing with traffic and aircraft noise, and even in conditions of changing thermal or other stress that can also be said to have an objective information content. I suggest that the study of subjective probabilities, i.e. of the information content that a stress situation has *inside* the black box, will prove to be important in moderate stress research. It may prove such an important dimension that level of intensity will be seen to be somewhat irrelevant in many cases. I will gives examples of such cases below when dealing with variable noise and temperature research.

3.2. *Time-Invariant Factors Independent of Stress and Strain*

Intelligence and personality are the most usual names given to such empirically deduced factors. It has been possible to ignore their effects in most extreme stress experiments by being very selective in the choice of subjects—fit young men, homogeneous groups of students or military personnel. The influence of these factors is probably small when placing limits on stress by means of internal sub-system criteria, even for more heterogeneous groups, but in moderate stress, using criteria of strain derived from other systems, they begin to play a part.

Bliding et al. (1966) have shown that personality can interact with heat stress to mediate performance, Johansson et al. (1970) that intelligence can interact with the effects of noise on performance, and Arvidsson et al. (1965) that personality is a factor in complaints of noise disturbance. One might object that ergonomists must take their experimental groups as they find them, but such studies could eventually be useful in job selection, and once the effects of personality and intelligence are better understood, the various diagnostic tests could prove useful in identifying high-risk groups in order to increase the sensitivity of experiments in the moderate stress region.

3.3. *Time-Variant Factors Closely Linked to Strain*

Arousal, attention and effort are the variables that fall into this category. Acquired skill at a particular task, and the ubiquitous fatigue, must also be placed here, if only to emphasize that they are more closely linked to strain than to stress.

However, fatigue as a concept is almost certainly redundant and comprises the combined effects of these time-variant factors acting together to mediate strain. It is worth pointing out that although level of skill is a factor linked to strain, task variables are properly parameters of stress and should be classed with the physical parameters of the environment.

In the space available, I cannot begin to review the enormous body of literature relevant to skill, arousal, attention and effort, but I would like to make two brief points. Firstly, that for military and for certain industrial purposes it has been appropriate to simplify the experimental situation in the extreme stress region by setting skill and effort to their maximum values and placing strict limits on the variations of arousal and attention by the design of the experiment. In this way, these four intervening variables have been effectively constants, permitting the derivation of stress–strain relationships with only minimal perturbation. In the moderate stress region, and under more normal working or resting conditions, these four intervening variables are free to vary independently, and greatly complicate the situation. It is for this reason that their study cannot be avoided in moderate stress research. Secondly, as a consequence of this, we require indices of each of them, indices which can at least indicate the direction in which they change, in order to decide between the many plausible interpretations of their operation that can be offered *a posteriori*. At present, the complexity of their interactions makes it impossible even to predict the direction of a stress-induced change in performance in the moderate stress region, as I shall illustrate below.

There are many indices of these four variables, and you pay your money and take your choice. For my money, the following seem to have good form. The technique of Evoked Cortical Response has been reviewed in this connection by Wilkinson (1967), who concluded that measures of arousal and attention can be derived by this means. Kalsbeek and Ettema (1965) have put forward sinus arrythmia as a measure of effort, and I have found that it is sensitive and useful in moderate stress research (Wyon 1968). Skill can be assessed as performance under standard conditions, but should be carefully distinguished from effort. The familiar results of Mackworth (1950) for telegraph operators under heat stress are probably an example of a case where effort, not skill, was the important factor. An easier version of the task might have proved too easy for the skilled operators and caused them to reduce the effort they exerted to such an extent that their performance was affected by the resultant effects of the heat on their arousal and attention. This is an example of the difficulties we are certain to find when we abandon the customary simplifications of extreme stress research and allow the intervening variables full play.

Table 1 sets out schematically the variables I have discussed. It can be seen that even this tentative list includes eight independently variable intervening variables, none of them accessible for direct measurement. In the following sections I will describe a number of results obtained for moderate stress, results which have been used to derive the above scheme and whose interpretation has been greatly influenced by it.

Table 1. The parameters of moderate stress research. (For explanation see text.)

Parameters of environmental stress		Thermal, visual, acoustic, olfactory, etc. (Task)
Intervening variables	Time-variant stress-linked	Subjective level of intensity Subjective probabilities. (Information content)
	Time-invariant, independent of stress and strain	Intelligence Personality
	Time-variant strain-linked	Effort Attention Arousal Skill
Parameters of strain	Internal Individual Total	Physiological Behavioural, subjective reports Performance

4. The Effects of Moderate Heat Stress on Children

In 1967, as part of a general design study of Swedish schools (Antoni 1969), it became necessary to set a maximum on permitted classroom temperatures. It was considered that neither subjective complaints nor altered patterns of behaviour would carry sufficient weight in resisting the powerful economic reasons for permitting classroom temperatures to rise on warm days. The available literature provided no grounds for believing that performance would be affected below air temperatures of 36°C, as found by Viteles and Smith (1966) (see reviews by, e.g., Pepler 1963, Wing 1965), yet it is a commonplace among teachers that school performance does deteriorate on even moderately warm days. From the analysis set out above, the reasons were postulated to be that effort and arousal are altered to sub-optimal values by the moderate heat. Such changes need not occur in conditions of high motivation to work, such as had prevailed in the published experiments. A new series of experiments was therefore carried out, under conditions as close to those of normal school work as possible, in all cases without disclosing the nature of the experiment to the children, and in some cases without them being aware that an experiment was in progress at all. These experiments have been described by Holmberg and Wyon (1967), Wyon (1969 a, b), and Ryd and Wyon (1970), and a brief summary of the results will suffice here.

In a test of learning carried out in a language laboratory, it was shown that oral performance significantly deteriorated at an air temperature of 27° as opposed to 20°C, and that the effect appeared to be confined to the less able section of the class (Ryd and Wyon *op. cit.*). The 13-year-old children were unaware that they were taking part in an experiment, and as questions and answers were in dialogue with a tape recorder, the effect of the teacher was peripheral and uniform.

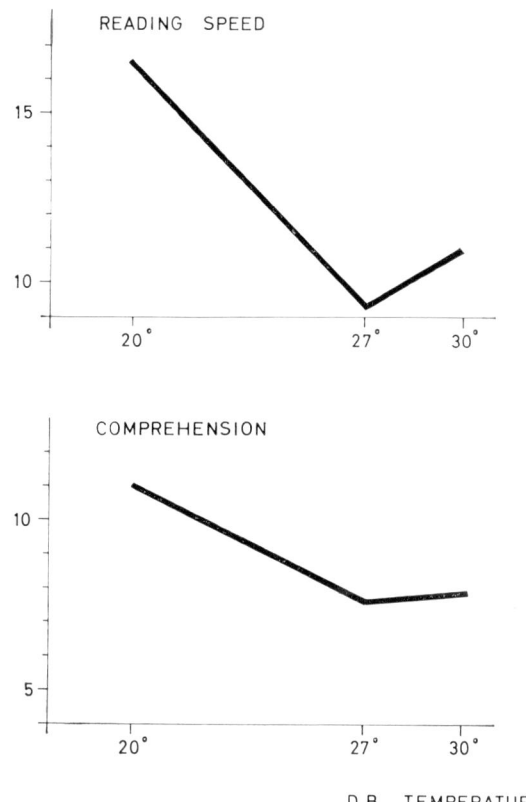

Figure 1. Data from Holmberg and Wyon (1967) showing reversal of temperature effect at 27°C.

In two further experiments (Holmberg and Wyon *op. cit.*), classes of 9-year-old children were exposed to air temperatures of 20°, 27° and 30°C in a balanced design, and classes of 11-year-old children to temperatures of 20° and 30°C. They carried out standard tests in the context of a normal lesson, supervised by their usual class teacher. Significant effects of temperature on performance were observed for reading speed and reading comprehension, and for certain other tests. In both groups, reading speed and comprehension deteriorated significantly at 30° as opposed to 20°C. The deterioration was of the order of 30 per cent. The effect of 27°C was also to cause significant decrement of performance, and in both cases the effect was greater than that of 30°C. Taking several tests together, the occurrence of this pattern was significant. Figure 1 shows the result obtained for the younger children. It was possible to examine also the effect of time of day for the older children, and Figure 2 shows the result obtained. The effect of temperature was more marked in the afternoon. This interaction was significant for reading speed, and shows also for comprehension.

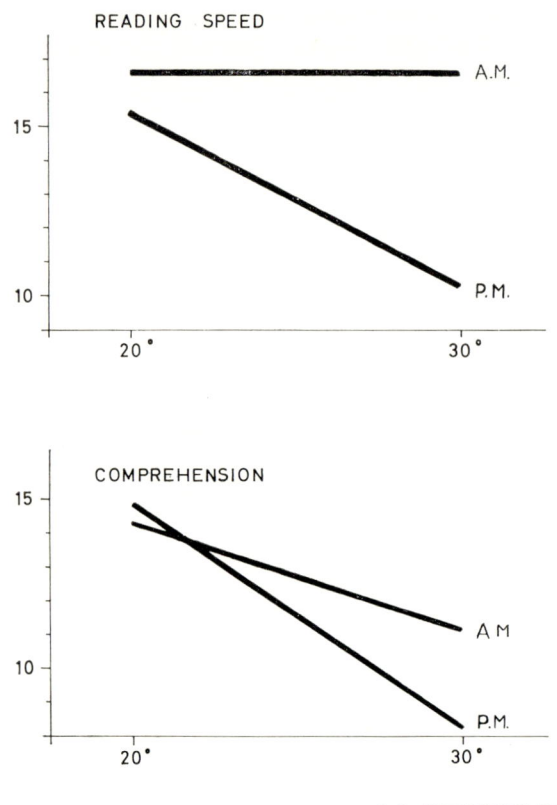

Figure 2. Data from Holmberg and Wyon (1967) showing interaction of time of day with temperature effect.

Observations of behaviour were made at the same time by observers concealed behind one-way mirrors in the specially-built classroom. They showed that significant postural changes took place successively with increasing temperature, such that the surface area of the body available for loss of heat was successively increased. Removal of clothing and observed sweating and vasodilation were also significantly and successively affected by the temperature changes, showing no sign of a minimum at 27°C. These observation results will be published shortly together with the results of observations made in the same way of children under imposed noise stress.

In a further field study, Ryd showed that a simple multiplication test was performed less well at air temperatures of 25° and 27° than at 20° and 23°C (Ryd and Wyon, *op. cit.*). Again, the least able of these 13-year-old children were more affected by the heat. Thus all of these experiments, using simple tasks very similar to ordinary school work, did show quite marked effects due to temperatures that occur quite frequently in occupied classrooms, even in Sweden. It is worth noting that in a further experiment of the same series, using 17-year-old children and ordinary examination questions, I failed to show any effect of temperature. It appears to be necessary to test quite specific abilities if a clear-cut effect of temperature is to be demonstrated. It is plausible that different abilities may be differently affected by temperature, even to the extent of being affected in opposite directions, and if this is so the total effect may be zero for a test whose performance depends on too many of them.

Provins (1966) has suggested that the effect of moderate heat stress is to lower the level of arousal, while higher levels of heat stress tend to increase the level of arousal. Taken together with the inverted-U relationship generally accepted to exist between arousal and performance, this non-inverted-U relationship between temperature and arousal could explain our results. A temperature of 27°C induced a lowered level of arousal, which was sub-optimal for the performance of our tasks. The same results could arise if a temperature of 27°C *raised* arousal to a value too high for our task, and indeed practically any results at all could quite easily be 'explained' by this complex interpretation. In the absence of some index as to the direction of change of intervening variables like arousal, such interpretations are fruitless. However, as part of the same schools design study, a climate chamber experiment was carried out (Johansson and Löfstedt 1969). It used the same levels of stress, achieved for unclothed children by subjecting them to temperatures of 30°, 36° and 41°C. Similar effects of temperature on performance were observed, and the key to a meaningful interpretation in terms of arousal theory was provided by two further tests. The Tsai–Partington test is known to be adversely affected by a high level of arousal (Eysenck and Willett 1962). In this experiment it was performed better at the intermediate temperature than at either of the two other temperatures, indicating a lowered level of arousal. In an unprepared auditive startle test, the children reacted to a lower signal level at the intermediate temperature, i.e. they were concentrating less on the primary task at this temperature, again suggesting lowered arousal.

Pepler and Warner (1968) have also reported a reversal of the temperature effect on performance (Figure 3). It occurred at precisely the same air temperature, 27°C, as for the Swedish schoolchildren. American students were shown to work more slowly through a programmed text at this temperature than at temperatures 3° and 6°C above or below. This is what we found for reading speed and comprehension. However, the American subjects were wearing only 0.5 clo, instead of the c. 1 clo in our field experiments, and they were optimally comfortable at 27°C. They also estimated that they exerted least effort at this temperature, which accords well with their slower performance. Percentage errors were not affected. Surprisingly, Pepler and Warner concluded that 27°C was the optimum temperature, reasoning that they worked with effortless efficiency at this temperature, and were comfortable. The question arises—optimum for whom? For the individual subjects, perhaps not very highly motivated in an experiment of six three-hour exposures, the optimum was quite possibly that temperature at which they could comfortably relax and be comfortable. An employer might take a different view, reasoning that the effort they reported was merely the effort they felt inclined to exert, not the effort required to maintain performance, because they did not in fact maintain their performance at the comfortable temperatures. The optimum temperature for the total system and for an employer would surely be 20°C. Subjective estimates of effort must be interpreted with care, and optimum temperatures must be defined with reference to the appropriate system level.

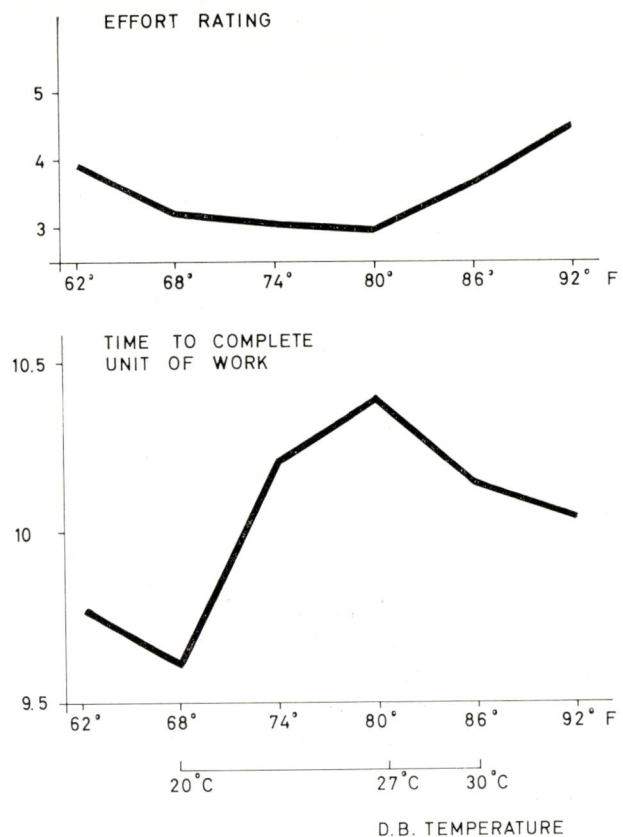

Figure 3. Data from Pepler and Warner (1968) showing reversal of temperature effect at 27°C.

In 1968 I carried out a further study to investigate more fully these factors of arousal and effort as affected by moderate heat stress (Wyon 1968, 1969 c). Forty-eight 11-year-old English schoolboys, normally clothed, worked in a climate chamber for two-hour periods in groups of four. One of their tasks was again a test of reading speed and comprehension. Unlike their Swedish counterparts in the field experiments, these boys were highly motivated, being volunteers with only nominal pay attending in their school holidays. They maintained their performance at 27°C at the same level as at 20° or 23·5°C. During the task, ECG records were recorded on a four-channel FM tape-recorder for a future analysis in which successive cardiac intervals were transformed to digital values and processed by computer program. A measure of sinus arrythmia was derived in this way (the SINAR score), and indicated that they had exerted significantly more effort at the higher temperatures. Three versions of the

Tsai–Partington test (Ammons 1955) were administered immediately after the reading test. On all three, performance was highly significantly better at 27° than at 20° or 23·5°C. In accordance with the general theory of Easterbrook (1959), cue-utilization is expected to be greater at lowered levels of arousal, and this would lead to improved performance of the Tsai–Partington test should 27°C have induced such a lowered level of arousal. A test of creativity or ' open-ended thinking ' was also performed by the boys, and there was an indication that their scores improved at 27°C, as would be expected if this type of thinking involves increased cue-utilization and a lowered level of arousal for optimum performance. It was found that at 27°C there was a highly significant increase in the number of repeated or redundant answers in this test, indicating a lowered criterion of what they considered to be a satisfactory performance. There was no corresponding increase in the total number of answers.

Future studies of performance under moderate stress could with advantage pay close attention both to measures of the ' cost of work ' and to subtle alterations in the strategy of performance. In an experiment to study slowly rising temperatures, carried out in Denmark last year, I have again used SINAR scoring as an indication of effort, and by using tests that can be analysed in terms of the signal detection theory of Tanner and Swets (1954) I have a means of following changes of strategy. These results are not yet available. The intention is to investigate whether slowly rising temperatures have an even more marked effect on performance than steady temperatures, by reason of the subject's interpretation of the situation. He may respond behaviourally to the predicted rather than to the actual temperature, decreasing effort and lowering his level of arousal *in advance* in order to combat the expected effects. Reduced muscular tension would reduce body heat production, and this adaptive behavioural response to moderate heat stress might be sufficient to induce the lowered level of arousal that we have so consistently observed. If such a change occurred predictively, even temperature could be said to possess an information content dimension as well as a level of intensity dimension.

5. The Effects of Moderate Noise Stress on Children

As indicated above (Section 3.1), I expect the information content of a noise background to be so influential that the level of intensity becomes almost irrelevant at moderate stress levels. This expectation is based on an analysis of how we deal with noise distraction. I recently put forward a brief outline of this in a conference paper (Wyon 1970), but as it is to be published in Swedish it perhaps bears repetition here in connection with the present discussion. Noise, or unwanted information through the ear, should be regarded as being actively processed between the ears in four stages—prediction, identification, interpretation and suppression. Together they constitute the mental load imposed by noise distraction. Firstly, on the basis of his experience and understanding of the noise climate, the subject *predicts* distraction by making subjective estimates of the probability of occurrence of the various possible noise signals, and of their probable nature. By a learning process, these estimates will gradually come to correspond to the objective probabilities, but if the noise climate is complex, unfamiliar or puzzling, the subject's expectations will correspond less well to reality and he will be more often surprised and distracted. At the occurrence of each noise signal, the subject must *identify* its source and location. His subsequent *interpretation* will depend on the associations he makes between it and his past experience, and the conclusions he then draws. Signals with a high meaning content for him personally are likely to be more distracting. Having identified and interpreted the signal, the subject is in a position to *suppress* it. He may do so very rapidly, as in the case of signals that are irrelevant and uninteresting to him, or he may be totally unable or even unwilling to suppress it at all.

Thus dealing with noise distraction in the moderate stress region can be viewed as a skill to be learned and subsequently carried out as a secondary task. By studying the course of ' learning to ignore distraction ' it should be possible to confirm whether the processes suggested above do occur. Regarding them as a secondary task makes it much easier to predict and interpret complex effects of variable noise in the light of the intervening variables scheme set out above (Section 3). I would expect that the intervening variables most relevant to the prediction stage of the task would be subjective level of intensity and subjective information content. Arousal and attention are likely to be the most important mediators at the stage of identification. McDonald *et al.* (1964) showed that alert subjects could learn to ignore noise distraction better than drowsy subjects. Attention is surely important by definition. Intelligence and personality play an important part in interpretation, while effort and the level of skill achieved, at both primary task and secondary noise task, are probably the most important at the stage of suppression. I do not suggest that these are the only intervening variables operative at each stage, but empirical investigation might profitably begin according to this scheme. Interpretation in terms of more theoretical models of noise perception, such as that set out by Leipp (1969), must be preceded by such empirical studies of behaviour in noise climates of widely different characteristics.

In order to see how an unfamiliar and unpredictable noise climate affected school work, we recently carried out a field experiment very similar to the heat experiments already described (Section 4). A total of one-hundred-and-ten 12-year-old children worked for 80 minutes in four class groups. Two of the classes were exposed to the noise climate under investigation, and two were taken as controls. The noise climate consisted of short bursts of white noise, ranging in intensity from 55–78 dBA. They were of random length, occurred in random sequence and with pauses of random length between them. They were audible, but their sound pressure levels were below the variable noise background of the class, which was present also for the control groups. Observations of individual children made from behind one-way mirrors showed that significantly more children appeared to be ' working with an obvious effort ' in the noise, and significantly more were ' distracted by others and stopped working '.

A commentary on the behaviour of the group as a whole has not yet been analysed, and is intended to show whether a chain reaction of disturbance spreads through the class. It would have been preferable to film behaviour so that observations could be made ' blind ' and repeated if necessary. Significant performance decrements due to noise were observed. In a self-paced numerical inspection task, they worked more slowly but were more accurate, possibly because they were repeatedly interrupted in the course of an inspection and had to start again. In a test of recognition memory, the noise-exposed classes were significantly less confident and less consistent in their answers. These preliminary results were reported by Wyon (1970) and it is intended to publish a full report when the analysis is completed. The experiment represents the beginning of a series to investigate the information content dimension of variable noise. It is hoped to demonstrate eventually that the measure of noise given by a noise meter has as little relevance to the distracting nature of a noise climate as the readings of a light meter have to the quality of a television programme. It is true that a light meter is useful to prevent damage and discomfort to the eyes of the viewer, in the same way that a noise meter is useful to prevent damage and discomfort to the ears, but audience research relates the subjective reports of viewers to more meaningful parameters than the readings of a light meter placed before the screen, and noise climate research should certainly not be content merely to relate subjective reports of noise to the readings of a noise meter.

6. Conclusion

The approach to moderate stress research that I have put forward in this paper is a highly personal one, representing the way in which I find it useful to structure the field. It is obviously tentative and subject to revision in many respects, but it does at least suggest a great deal of research. I hope that it may stimulate somebody to do the necessary experiments, even if only to prove my interpretations wrong. Studies must be made in parallel on all types of moderate stress, and it is often extremely difficult to see the relevance of results obtained in one speciality to results obtained in another. By uniting in a common aim to demonstrate that my scheme does *not* apply to all types of moderate stress, I hope that specialists in many fields may be encouraged to compare notes.

Dans cette étude consacrée à la contrainte modérée, l'auteur s'interroge sur les principes qui guident le choix des critères de contrainte et d'astreinte. Les paramètres les plus importants dans ce type de recherche sont identifiés et placés dans le contexte d'un schéme empirique en tenant compte des possibilités d'interprétation et de leur pertinence.

En guise d'illustration, l'auteur cite une recherche qu'il a effectuée sur des enfants soumis à une contrainte thermique et dont il interprète les résultats en terme de niveau d'éveil et d'effort. Une approche est suggérée pour l'étude des effets du bruit dans la zone de contrainte modérée. Mais il ne semble pas que dans ce genre d'étude le niveau de pression sonore soit le paramètre adéquat; c'est du moins ce qui ressort d'une recherche préliminaire sur les effets des bruits intermittents de faible intensité sur les enfants.

Die Untersuchung " Mässiger Belastung " wird als ein in sich geschlossenes Forschungsgebiet herausgestellt und das darin geltende Prinzip der Auswahl von Kriterien für Belastung und Anstrengung diskutiert. Die wichtigsten Parameter " Mässiger Belastung " werden bezeichnet und in einem empirischen Schema zusammengestellt, interpretiert und diskutiert. Untersuchungen des Autors an Kindern unter Hitzebelastung werden benutzt, das Schema zu erläutern (in Werten der Erregung und Anstrengung). Eine Probeuntersuchung von Lärm in der Zone " Mässiger Belastung " wird angeregt, in der Schalldruckmessungen wenig Erfolg versprechen. Diese Ansicht wird durch vorläufige Resultate einer Untersuchung an Kindern bei intermittierendem mässigen Lärm gestützt.

References

Ammons, C. H., 1955, Task for study of perceptual learning and performance variables. *Perceptual and Motor Skills*, **5**, 11–14.
Antoni, N., 1969, Projekteringsunderlag för skolbyggnader för grundskolan. (A design study of school buildings.) *Report 50/69, Swedish National Institute for Building Research* (**17** booklets, SKr. 300: AB Svensk Byggtjänst, Box 1403. 111 84 Stockholm).

ARVIDSSON, O., JOHANSSON, C. R., OLSSON, K., O. WIGEMAN, H., 1965, Samhällsbuller—en sociologisk-psykologisk studie. (A sociological and psychological study of noise in society.) *Nord. Hyg. Tid.*, **XLVII**, 153–186.

BLIDING, G., HAGBERG, B., LÖFSTEDT, B., u. TRYGG, L., 1966, Mental prestationsförmåga i hög värme. (Mental performance at high levels of heat stress.) *Nord. Hyg. Tid.*, **XLVII**, 1.

EASTERBROOK, J. A., 1959, The effect of emotion on cue-utilization and the organization of behaviour. *Psychological Review*, **66**, 183–201.

EYSENCK, H. J., and WILLETT, R. A., 1962, Cue-utilization as a function of drive: an experimental study. *Perceptual and Motor Skills*, **15**, 229–230.

HOLMBERG, I., O. WYON, D. P., 1967, Skolprestationernas beroende av temperaturen i klassrummet. *Pedagogisk-psykologiska problem*, Nr. 55, 44 p. (English language version, 1969, The dependence of performance in school on classroom temperature. *Educational and Psychological Interactions*, Nr. 31, 20 p. School of Education, 200 45 Malmö 23, Sweden.)

KALSBEEK, J. W. H. and ETTEMA, J. H., 1965, Sinus arrythmia and the measurement of mental load. *Paper at British Psychological Society Conference, London.*

JOHANSSON, C. R., BENGTSSON, K., BÅÅTH, I., O. CRONA, G., 1970, Skolprestationens beroende av bullerstörning med särskild hänsyn till intelligens och personlighetsfaktorers betydelse för prestationspåverkan. (The effects of noise on school performance and the interaction of intelligence and personality with these effects.) *Unpublished report, Department of Psychology, University of Lund, Sweden.*

JOHANSSON, C. R., O. LÖFSTEDT, B., 1969, Klassrumstemperaturens inverkan på skolprestationen: Laboratorieförsök i klimatkammare. (The effects of classroom temperatures on school performance: a climate chamber experiment.) *Nord. Hyg. Tid.*, **XLX**, 9–19.

LEIPP, E., 1969, Les mesures physiques du bruit et leur signification du point de vue de la perception. *Cahier du Centre Scientifique et Technique du Bâtiment*, Nr. 869, 6 p. (4 Avenue du Recteur Poincaré, Paris XVIe).

McDONALD, D. G., JOHNSON, L. C., and HORD, D. J., 1964, Habituation of the orienting response in alert and drowsy subjects. *Psychophysiology*, **1**, 163–173.

MACKWORTH, N. H., 1950, Researches on the measurement of human performance. *Medical Research Committee Special Report* No. 268 (London: H.M.S.O.).

MACPHERSON, R. K., 1962, The assessment of the thermal environment: a review. *British Journal of Industrial Medicine*, **19**, 151–164.

PEPLER, R. D., 1963, Performance and well-being in heat. In *Temperature: Its Measurement and Control in Science and Industry.* Vol. 3 (Edited by J. HARDY) (New York: REINHOLD), pp. 319–336.

PEPLER, R. D., and WARNER, R. E., 1968, Temperature and learning: an experimental study. *ASHRAE Trans.*, **74**, 211–219.

PROVINS, K. A., 1966, Environmental heat, body temperature and behaviour: an hypothesis. *Australian Journal of Psychology*, **18**, 118–129.

RYD, H., and WYON, D. P., 1970, Methods of evaluating human stress due to climate. *National Swedish Institute for Building Research*, Nr. D6/70.

TANNER, W. P., and SWETS, J. A., 1954, A decision-making theory of visual detection. *Psychological Review*, **61**, 401–409.

TAYLOR, M. M., 1967, Detectability theory and the interpretation of vigilance data. *Acta Psychologica*, **27**, 390–399.

VITELES, M. S., and SMITH, K. R., 1946, An experimental investigation of the effect of change of atmospheric conditions and noise on performance. *ASHVE Trans.*, **52**, 167–182.

WILKINSON, R. T., 1967, Evoked response and reaction time. *Acta Psychologica*, **27**, 235–245.

WING, J. F., 1965, A review of the effects of high ambient temperature on mental performance. *AMRL–TR–65–102*, vi, 34 pp.

WYON, D. P., 1968, The effects of moderate heat stress on the mental performance of children. *MSc. Dissertation in Applied Psychology, University of Aston, Birmingham.* (Published as WYON, D. P. (1969 c).)

WYON, D. P., 1969 a, The effects of classroom temperatures on school performance: studies in the field. *Nord. Hyg. Tid.*, **XLX**, 20–23.

WYON, D. P., 1969 b, The effects of temperature on arousal and effort in mental work. *Proceedings of the Fifth International Biometeorological Congress, Montreux, Switzerland.* (Abstract only.)

WYON, D. P., 1969 c, The effects of moderate heat stress on the mental performance of children. *National Swedish Institute for Building Research*, Nr. D8/69, 83 p.

WYON, D. P., 1970, Performance and behaviour of schoolchildren during low-level but intermittent noise. *Proceedings of the Noise Climate Conference, March 12, Stockholm. Building Research Council.* Report No. R.36/70, 12–15 (In Swedish).

WYON, D. P., LIDWELL, O. M., and WILLIAMS, R. E. O., 1968, Thermal comfort during surgical operations. *Journal of Hygiene*, **66**, 229–248.

Emotional Illness and the Working Environment

By J. F. L. Aldridge

IBM United Kingdom Limited, London, England

Some of the difficulties associated with the measurement of alterations in mental health resulting from the working environment are discussed. Causes of stress are described which stem from the individual and his relationships, and from the organization of which he is a part. The need for a re-evaluation of the industrial physician's role and for future collaborative studies of occupational mental health is suggested.

1. Introduction

Industrial medicine has in the past tended to concentrate its efforts on the physical and toxic hazards of occupations and has spent surprisingly little time and energy on an evaluation of the psychological environment. Considering the disabling and sometimes fatal effects of the former the priorities that were allotted are not altogether surprising, although there is not a great deal of evidence that this situation is changing very greatly in the United Kingdom. There can be no doubt that on the one hand the control of the physical and chemical environment has improved enormously, while on the other hand it has become fairly generally accepted that minor psychological illnesses have steadily increased and that the effects of these illnesses account for a considerable proportion of cases seen in the consulting rooms of general practitioners and of the industrial physicians.

Russell Fraser (1947) in an important survey showed that 30 per cent of 3000 workers in the engineering factories studied suffered from some form of neurosis during the course of six months; and the Report of the Department of Health and Social Security for 1968 (1969) shows that the sum of incapacity for men suffering from mental, psychoneurotic and personality disorders, nervousness, debility and headache accounted for 22·8 million days. This places mental illness in the second place after bronchitis (30·9 million days) and before accidents (22·3 million days) and heart disease (19·2 million days).

From this, one might reasonably conclude that some industrial doctors should be liberated from what could loosely be called the mechanical side of occupational medicine in order to allow more time to deal with the emotional aspects of the environment—a field which was well reviewed by McLean (1966). Whether this will actually occur remains to be seen, but there should be no doubt that the mental health of an organization is of paramount importance to its success and needs all available manpower to assess the size of the problem and to develop techniques for its improvement.

2. Measurement Difficulties

An objective scientific assessment of the quantity and quality of mental illness that occurs in the working environment is extremely difficult and it soon becomes clear that there are many factors in doctor, patient and society which affect the final answer: the doctor in his surgery makes his own judgment as to whether symptoms are primarily due to organic or to emotional causes, and while one doctor may treat symptoms simply at their organic face value, another may search successfully for an underlying psychological cause. Not only is this variation of perceptive ability different from one doctor to another but it is also variable from time to time in each individual doctor. Doctors within industry as well as those outside become known for their sympathy, or lack of it, when dealing with emotional illness, and frequently the patient will try to adopt an approach which he believes will be acceptable to the doctor he consults, or alternatively chooses a doctor who by reputation is receptive and has been helpful to patients with similar illnesses. Just as many doctors may feel more comfortable treating organic disease, so some patients find that the expression of their own feelings can be frightening and they often prefer to clothe their illness with somatic symptoms which are commonplace and acceptable to the culture in which we live.

Measures of absenteeism can be most helpful in following broad trends of sickness behaviour in various working groups but are notoriously inaccurate when it comes to the diagnostic cause of the absence. The possibility of the patient using somatic symptoms to present his emotional illness, and of the doctor who dismisses emotional symptoms or interprets them in terms of organic disease, have already been mentioned and will clearly affect the accuracy of recording. Additionally, the doctor may, with the best of intentions, disguise a psychiatric diagnosis to protect his patient from stigma. Although attitudes to psychiatric illness are changing slowly for the better, there are still those who inflexibly assume that anyone who has suffered from some form of mental illness is no longer fit to take a responsible place within industry.

Short uncertificated absences of three days or less are frequently unjustifiable on objective clinical grounds and are probably more a reflection of motivational influences than of significant organic illness (Taylor 1968). These are in many cases the method by which the individual may temporarily remove himself from the tensions of his work-place: not only can symptoms which would normally be entirely bearable become significant as a result of difficult relationships or experiences at work, but also accidents may be used as socially acceptable reasons to justify an absence (Hill and Trist 1962). The industrial arbiter of the short uncertificated absence is generally the absentee's immediate manager whose attitude is shaped by the previous work performance and absence record of the individual concerned and his own ability to relate cause with effect. There can be no doubt that trained and perceptive management in liaison with a doctor can frequently assist greatly in case-finding and in the resolution of quite long-standing emotional difficulties.

By no means all of those who develop emotional illness at work will, or should, report to the industrial physician but will rather go direct to their own general practitioner from whom they will expect some appropriate form of treatment. Communications between industrial physician and general practitioner are rarely good, with the consequence that each one sees only a part of the total picture both as regards the community and the individual.

Regrettably, communications are frequently even more difficult in cases of mental illness due to the attitudes of the doctors concerned, with the result that the least information is available in an area where most is needed.

The industrial physician whose population is scattered in small locations over a wide area is put on an even greater disadvantage and has to rely upon management, local general practitioners and the individuals themselves to bring cases to his attention.

Many forms of behaviour which reflect emotional distress do not involve either absence from work or visits to the doctor and are frequently accepted as being entirely within the bounds of normality. At their simplest these may be a transient sleep difficulty or a temporary increase in cigarette consumption and more seriously may present themselves as an abnormal use of alcohol or as a deterioration in family relationships at home. There are cultural restraints on the amount of overt anger or aggression that can be expressed and is acceptable in a working environment, and wives and children may be the recipients of aggression that has been suppressed during the day in frustrating circumstances. Marital breakdown and infidelity may also, in some cases, be measures of excessive tensions at work as may aggressive and foolish driving on the roads, disruptive behaviour at work and unusual alterations of work performance.

From the foregoing it should be readily appreciated that the ways in which emotional work difficulties may manifest themselves are diffuse and may be confusing. In addition, it is frequently difficult to separate the social or domestic factors from the industrial factors in the causes of emotional illness, and significant areas of a patients' history may be denied to either the general practitioner or the industrial physician because the patient does not see it as relevant to the doctor in question, having placed each doctor in a well-defined role without much overlap.

3. Some Causes of Emotional Ill Health

The work-related causes of emotional illness are equally as diffuse as the effects and may be separated, somewhat artificially, into those causes which stem principally from the organization in which the individual works and those which arise as a result of personal or interpersonal difficulties within that organization. The factors which are discussed below are those which seem to be specially relevant and important for the managerial and professional sectors of business rather than from the whole of industry.

It must be clear that there is seldom only one factor involved at any one time. Each individual's ability to cope with stress varies with the events and his experiences in all the various reference groups and environments of which he is a part. It would be completely unrealistic in normal circumstances to view a man solely in relation to his job or solely in relation to his family, although in many instances it is possible to discern the principal cause.

The success of an organization must depend to a large extent on its ability to recruit those whose skills, intellect and personality can be integrated effectively within that organization. Measures of skills and aptitude, intelligence and personality are available in many forms and are sometimes used as part of the evaluation of a candidate for specified employment. While with aptitude and intelligence testing it is possible to gain fairly accurate information, the impression gained from personality testing in these circumstances may be totally misleading with the candidate answering the questions in the way in which he believes will best please the interviewer. While the results of a personality test may be used as a talking point in a subsequent interview, it is the interviews in themselves that are usually the most important method of selection—i.e. the impression made by the candidate upon his interviewer. Two images of every company are projected to the outside world: the first is what the company wishes to project about its overall philosophies in regard to its management, its employees, its products and its commitment to the social community; and the second is that which derives from past and present employees and which inevitably includes both fantasy and folk-lore. A candidate who is keen to obtain a new position assesses these images in relation to himself and judges, by what he knows, whether his own view of himself is likely to be compatible. If he decides that there is reasonable compatibility and the company agrees by calling him for interview, there follows a mutual appraisal with an interviewer representing the company and with the candidate representing himself—both playing roles to a variable extent to present themselves in a favourable light—from which a highly subjective judgment is made. Although practical studies to test the extent to which this system is successful are difficult if they are to include candidates who are rejected, Gellerman (1968) advocates the use of a continuous review and evaluation of selection performance.

That the system is as successful as it seems to be is due partly to the large amount of self-selection that is involved and partly to the flexibility of the younger entrants.

The mistakes that can arise are illustrated by a 38-year-old man whose job included the requirement of making frequent new contacts outside his own company. Although he had considerable experience in other positions he developed feelings of increasing apprehension and anxiety with the result that he rapidly became totally ineffective and ceased to make a contribution in his job. It then transpired that he had always had difficulties in making new relationships. His personality was judged to be basically incompatible with the job for which he was selected and after a transfer to a staff support role he improved rapidly and completely.

The use of measurable objectives which can be set, reviewed and evaluated in terms of achievements at some future date is well known as a management technique and when properly administered can provide a sensible system for individuals and for groups (Drukker 1955). The results will depend very largely on the abilities of the individuals concerned to set valid, realistic objectives and on a commitment to flexibility during the period of time under consideration. Unfortunately, subsequent planning frequently outweighs original objectives with the result that individuals are striving to reach their goals with reduced resources of people and financial support and without the reviews which would take the altered circumstances into consideration. Initially, by allotting stringent priorities to problems, decisions and action, it is possible to keep pace with the demands of increased productivity, but the point may come when individual work-load is perceived as being intolerable when related to the frustrations of obtaining the information necessary to manage, with resulting ill-health.

All too often there is insufficient definition of responsibility and authority attached to management and professional tasks with the result that confusion and misunderstanding can occur over such a basic process as the making of decisions. It is easy in this situation for each level of management to refer a relatively unimportant issue to a higher level, or alternatively for someone at a low level to make an important policy decision which should have merited discussion by senior management. It is this lack of precise identification within the structure of the organization which underlines feelings of insecurity.

All working groups tend to develop their own norms regarding such things as their speed of work, the amount of work that is accomplished in a day and the time at which work is finished at the end of the day. These norms will be partly formal and partly informal and will depend very largely on the culture of the total organization of which the particular group is a part. The informal norms may cause tensions by the pressures that are placed on the individual to conform, even though there may be no logical need to conform from the viewpoint of efficiency and productivity. The behaviour becomes habitual and unnecessary rather than selective and useful and is well illustrated by the department who need to be seen to stay late every evening to ' prove ' their persistent effort. The same culture may question whether the man who leaves on time, and may therefore be well organized, has too little to do.

Although the eventual aims of all groups of an enterprise should be identical, it is obvious that some of these groups will inevitably be in conflict with each other. For example, it would not be reasonable to expect Sales Management to have the same viewpoint or priorities as Manufacturing Management and yet both groups rely heavily upon the abilities of the other. Many would say that a Personnel department should constantly be in conflict with the Finance department on the grounds that the best personnel policies will be created in this way, but the danger is that the conflict may become unhealthy and unproductive. Unless there is sufficient machinery for communication so that the conflict can be fully exposed and openly discussed, there is an habitual tendency for each side to become increasingly resistant to change and interference and attitudes develop which hinder the working relationship.

The patterns of communication, both formal and informal, that exist in any organization should not only facilitate the resolution of inter-group conflicts but also provide a system for passing accurate and relevant information as quickly as possible under the particular circumstances. Frequently, however, the design seems to inhibit useful exchanges and it can appear as if the object is to avoid bringing conflict into the open in case it turns out to be too destructive. Equally important are difficulties in obtaining information when the communication system is poorly developed or is overtaken by the growth or complexity of the organization and when the need for change is not acted upon sufficiently early.

Provision should be made to allow for close understanding between a man and his manager and may include formal, periodic meetings where the past may be reviewed critically, future plans and aspirations compared and guidance given to improve performance and capability. A system such as this should provide opportunities for the individual to examine his position and his progress, but nevertheless may lead to a conflict between the man's view of his own capabilities and promotability and that of his management. It may be difficult in early interviews to avoid giving an individual expectations which at a later stage become impracticable and unrealistic. This same conflict of ambition with reality may be entirely self-generated when the man realizes that he is no longer capable of advancement and that the ceiling he has reached is lower than the level to which he has always aspired.

As an example, a 39-year-old man was appointed to a company in a middle management position. He was given glowing references by his previous employers and seemed aggressively certain of his ability to succeed. Within a year of his starting work his department began to suffer from many difficulties, some due to market circumstances and some to his mismanagement. After a month or two, it became apparent that he was unable to deal with the situation and he was moved into a less exposed job in another location. Shortly afterwards his consumption of alcohol increased and six months later he developed symptoms of anxiety and depression.

It is often said that modern management needs to be flexible in its attitudes and it is certainly true that it needs to be flexible in the geographical sense. Frequent relocation of management and professional staff has become a normal part of the development of a successful career within industry. Unfortunately, the frequency of the moves and the haste with which they are so often required may be more likely to indicate poor forward planning and point to areas of inflexibility within the organization rather than suggest a stable, dynamic form of management.

Although a move is usually rewarded by an increase in salary and sometimes with an increase in status, it carries with it all the tiresome details and complications involved in buying and selling houses in addition to handing over one job and taking over another that is probably larger and more responsible. There is the need to disrupt one set of social ties within a community and to develop another, and the individual is also confronted with the conflicts of his roles at work and at home.

Psychiatric symptoms seem not uncommon in association with house moving, and in one case a 34-year-old man of high intelligence, engaged in intricate development work, developed increasing tension, insomnia, bad temper and occasional tearfulness. He was normally keen to accept all intellectual challenge at work as well as in his social environment, but found that the important project on which he was working came to a halt because he could not deal with the inherent frustrating difficulties. His symptoms arose during the two months preceding his move which had been surrounded by immense delays and disappointments.

It may be extremely difficult to equate the demands of a business enterprise in the role of an ambitious effective manager with that of a family that requires a husband and father to participate fully in those roles. The young man willingly gives his time and energy to his company and may be encouraged by a wife who is ambitious for his success and sees his advancement in real terms of his ability to provide for the home. In time, especially with the arrival of children, the wife may rightly see the need for a change in priorities and may begin to put pressures on her husband to play a greater part in giving emotional as well as practical support in the running of the household. It would be entirely wrong to blame organizational pressures and business needs for the majority of cases where the husband persists in giving the greater part of his attentions to his work. More commonly it is due to the man's own immense drive to compete and to succeed over all other. Although this behaviour can, of itself, be described as neurotic and although the achievement of each goal is accompanied by the setting of increasingly difficult targets, the results, in human terms, are likely to affect those in contact with the individual rather than the individual himself. The wife may react jealously to the enterprise which appears more attractive than herself and the marriage may, as a result, break up. Those who have this drive will often expect all others to have the same energy and the same high standards of work behaviour and there may be little understanding for deviations from these standards. Similarly, the manager who is unduly anxious may project his anxieties on to those around him with the result that tensions generally are heightened, and in a situation such as this it is not uncommon for the most vulnerable person in the group to become ill—as if he had been chosen by the group to express their tensions.

The great majority of people wish to complete their work to a standard that is satisfactory to themselves as well as to the requirements of their job. This may include wider involvement and interest than is strictly required by the job, but yet which is greatly important to individual motivation. Sacrificing these wider interests can produce much frustration, especially when the reason is lack of flexibility in day-to-day operations, with the result that short-term objectives take precedence over long term.

The general pace of life and the current urgency with which so much business is transacted have developed in parallel with the growth and proliferation of travel facilities and particularly of the commercial air routes. Frequent and lengthy travelling, which can become a tiring strain after the initial novelty has worn off, has become expected on the part of management. Pressures on time frequently dictate that the travelling is undertaken during leisure hours, and it is exceptional to find a company that pays more than lip service to the need for adequate relaxation and adaptation to new environments before embarking upon important meetings or other business. The business traveller usually crams as much as possible into his days and his programme of travel and meetings is compressed in order to minimize the impact upon the time he has available. In one documented case, a 46-year-old middle manager sought treatment for insomnia and anxiety which had persisted for several weeks following a particularly full and difficult transatlantic trip.

This management mobility can sometimes also affect the training and development at junior management levels: the absentee manager is no longer able to supervise and guide those who are responsible to him and by his absence increases individual work-load. While authority will have been delegated, so will responsibility, and the result can be extremely difficult for someone who lacks support on the one hand and experience on the other. This is illustrated by a 25-year-old man who developed recurrent nausea, vomiting and symptoms of depression within a few months of his promotion into management. His symptoms were partly associated with an increase in responsibility as he had immediately been faced with some difficult human problems; but he had also lost the support of his immediate superior who was for a large amount of the time transferred to a different location on a special project.

Similar problems can arise in those who have recently been promoted. It is often difficult for companies to balance their present need for technicians with their future need for management: to have too many with all the attributes of management potential breeds frustration and an increased turnover of staff if vacancies in management do not occur. Often, the growth of a company can place severe strains on its ability to select and train managers, with the result that anyone who shows success is immediately promoted and may well find himself pushed beyond his capabilities.

The step into management for the first time may be considerable in that it frequently involves a complete change of outlook and activity. Previous skills and technical ability are relinquished from the practical point of view and the new manager has to become an administrator and a leader virtually overnight. In the first instance, the rewards in terms of status and money may not obviously outweigh the increased responsibilities and the loss of direct technical contact in the previous job.

Management training and certain methods of management selection may in themselves be associated with emotional illness of varying degree. They are frequently run under high pressure with little or no time for relaxation or exercise, and working time may encroach upon the usual hours of sleep. The environmental setting and the inherent pressures often encourage overeating as well as an increased alcohol intake. Elements of the selection mechanisms for courses as well as the course contents are bound to be competitive even if this is not overt, and there may also be the realities (and the fantasies) of reports of individual performance to senior management at the conclusion of the course. The design and techniques that may be used to increase interpersonal and group skills are sometimes especially implicated in causing mental illness, and difficulties can certainly be caused by inadequately trained training staff and inappropriate training models. It is hard to say in many cases whether the emotional illness is a direct result of the training course or whether the course provides an opportunity and an excuse to be ill in a situation which is away from the normal working environment. It is quite possible that ill health occurring in association with management training may reflect the wider pattern of illness occurring in the everyday life of the enterprise.

4. Need for Scientific Study

These are only some of the factors which can affect mental health at work and their relative importance will vary with different observers and in different enterprises. Many other factors of a more subtle nature undoubtedly exist and there is a positive need to tackle and overcome these difficulties of measurement mentioned earlier to be able to arrive at a clear understanding of the total size of the problem. Information will have to be collated from doctors working within and without industry, from psychiatrists, behavioural scientists, management and, importantly, from employees.

There is already a need to re-examine the role of the industrial physician in the light of changing circumstances and also to reorientate much of his training. The future emphasis must be on people and an understanding of their behaviour to provide improved skills to assist with the planning, education and training required to develop a healthy organization and environment.

Les auteurs discutent des difficultés que l'on rencontre lorsque l'on tente d'évaluer les altérations de la santé mentale consécutives à l'environnement du travail. Ils décrivent des causes de stress relevant soit de l'individu et de ses relations, soit du groupe dont l'individu fait partie. Ils suggèrent enfin une révision de la conception du rôle du médecin du travail, et insistent sur la nécessité d'entreprendre des études multidisciplinaires sur le problème de la santé mentale en milieu professionnel.

Einige der Schwierigkeiten werden diskutiert, die mit den Änderungen des Geisteszustandes zusammenhängen, welche von der Arbeitsplatz-Umgebung verursacht werden. Belastungsursachen werden beschrieben, welche im Individuum selbst, in seiner Familie und in den Organisationen, denen er angehört, liegen. Es wird vorgeschlagen, die Rolle des Werksarztes neu zu bewerten und mit ihm zusammen künftig die berufliche Geistesgesundheit zu studieren.

References

DEPARTMENT OF HEALTH AND SOCIAL SECURITY, 1969, *Annual Report for* 1968 (London: H.M.S.O.).
DRUKKER, P. W., 1955, *The Practice of Management* (London: HEINEMANN).
FRASER, R., 1947, *The Incidence of Neurosis in Factory Workers* (London: H.M.S.O.).
GELLERMAN, S. W., 1968, *Management by Motivation* (NEW YORK: AMERICAN MANAGEMENT ASSOCIATION, INC.).
HILL, J. M., and TRIST, E. L., 1962, Industrial accidents, sickness and other absences. *Tavistock Pamphlet No. 4. Tavistock Publications.*
McLEAN, A. A., 1966, Occupational mental health: review of an emerging art. *American Journal of Psychiatry*, **122,** 961–976.
TAYLOR, P. J., 1968, Personal factors associated with sickness absence. *British Journal of Industrial Medicine*, **25,** 106–118.

Session 4
Panel Discussion

G. R. C. Atherley, University of Salford. The term 'physiological damage' used by Dr. Wyon is to my mind a contradiction in terms. 'Physiological' to me implies normal function, and is pleasant, whereas something pathological is exactly the opposite. Dr. Aldridge, when considering anxiety, points out that normal anxiety can pass into a pathological state. I would welcome a comment from our two speakers on what they consider to be the dividing line between physiology and pathology.

Wyon. In the fields of moderate stress, if we seek pathological changes as a criterion of strain, we will not find them. Pathological changes form no part of moderate stress. Where I used the term 'physiological strain', I meant to imply indices of stress susceptible to physiological methods of measurement, and which can be interpreted in physiological terms. These are parameters which, above a certain level, can be termed undesirable, but not impermissible. In considering sub-optimal conditions, pathological indices are unlikely to be of help.

Aldridge. I do not feel well qualified to answer this. Perhaps one should consider sub-optimal effects as transient, not seriously affecting the quality of life, whereas pathological processes may cause permanent and serious deprivation.

Murray. While bringing this session to a close, I would not pretend to summarize either this afternoon's proceedings or those of the last two days. I have much to think about, both from the point of view of industrial hygiene, and from that of ergonomics, and I think we all have much to think about in the coming months. One particular point that arises is that occupational medicine must spend more time studying the individual in relation to his environment, rather than as at present studying the environment in relation to individuals. This comes back full circle to Lord Robens' introductory remarks.

May I thank all our speakers and discussants and bring the conference to a close.

Subject Index

A

Absenteeism, 84
Acoustic stimuli, 6
    measurement of, 7
Adaptive response (to heat), 79
Adrenal medulla, 9, 14
Adreno cortico trophic hormone (ACTH), 9, 14
Aggression, 60, 85
Air ionization, 68
    movement, 71
    pollution, 49, 57
Alcohol, 52, 58, 65, 85, 90
Anaemia, 31, 34, 37
Ankle spats, 41
Anxiety, 9, 52, 92
    depression, 6, 14
Aptitude, 38
    test, 32
Architects, 69
Arousal, 9, 57–65, 68–82
    and performance, 77
Asbestosis, 49
Atmospheric pollution, 65
    variations, 39, 40, 49
Attention, 9, 54, 73, 80
Audiogram, 16, 20–23
Auditory flutter fusion threshold, 57–64
Autonomic nervous system, 9, 14

B

Bacteria, 40
Baths, 41
Behaviour, 68, 70

Bladder cancer, 49
Blood pressure, 46
    vessels, 40, 42
Body temperature, 68
Boilermakers, 20
Boredom, 66
Brain, 58
Bromidorosis, 40
Bronchitis, 83
Bursa, 43

C

Calcaneum, 43
Cancer, 49
Carbon monoxide, 49, 57–64, 65
Carboxyhaemoglobin, 57–64
    half life, 65
Card dealing and sorting, 29
Cerebral depressant drugs, 57–64
Cerebrum, 50
Children, 68–82
Chiropody, 39, 42–43, 45, 46
Chloral hydrate, 51
Chloroform, 51
Cigarettes, 63, 85
Circadian rhythms, 28–30, 46, 47, 76
    variation, 10
Clothing, 77
Code transcription, 29
Coffee, 58, 63
Cold, 40–41
Collapse, 68
Colour stimulus response test, 63
Comfort, 69, 70, 71
Cornell Medical Index Questionnaire, 54, 55

Corns, 39, 41
Critical flicker fusion threshold, 57–64
Cue utilization, 79
Cyanose, 41

D

Death certificates, 49
Depression, 52, 88–89
Dermatitis, 33
Dermatophytosis, 41
Discomfort, 70
Disease, 33, 67
Dizziness, 52
Doctors and mental illness, 83–84
Drink, 69, 85
Drivers, 63, 66, 85
Drugs, 57–64
Dryness, 40
Dust, 2

E

Ear-muffs, 59
Ears, 69
E.C.G., 78
Economic growth, 31
Effort, 68–82
Emotional illness, 83–91
    causes, 85–90
Engineers, 69
Environment, 83–91
Environmental stresses, 68–82
Eosinophils, 6, 9, 10
Epidemiological research, 49
Evaporation, 40
Evoked cortical response, 73
Exercise, 58, 65
"Expectancy" (vigilance experiments), 72
Eyes, 69

F

Family size, 36
Fatigue, 52, 73
Fatty-acids, 40

Feet, 39–45
Fitness, 34
Flat-foot, 42
Flatulence, 52
Flicker-fusion frequency, 29
Foot disorders and deformities, 39–45
Foot infections, 42–44
Footwear, 39–45
Forefoot splay, 42
Fresh air, 2
Fumes, 63
Fungi, 41, 44

G

Gases, 57–64

H

Haemoglobin, 32, 34, 37, 52
Haemorrhage, 44
Hallux valgus, 43
Headache, 52, 65, 83
Health, 28, 67, 83–91
    and productivity, 31–38
Hearing loss (noise induced), 16
Heart disease, 83
Heat, 5, 32, 39, 40, 70
    stress, 68–82
Heel, 43
Heron's Personality Questionnaire, 54, 55
Hot air, 1
House moving and psychiatric symptoms, 88
Humidity, 39, 40
Hypergranulation, 44
Hyperidrosis, 40, 41, 44

I

Illness, 31, 34, 37, 38, 83–91
Illumination, 68
Imposed noise, 68–82
Impotence, 65
Industrial Footwear, 42

Industrial medicine, 84, 90
Ingrowing nails, 44
Intelligence, 72, 80, 85
Intermittent noise, 68–82
Interviews, 86
Ionization-air, 68
Iron deficiency, 31, 34
Irritability, 14
Itching, 41

K

Ketosteroid-17 urinary, 6, 9, 10, 14, 65

L

Lead, 52
Learning, 65, 75, 79
Leather, 40
Letter stimulus response test, 63
Light, 57, 59
Lighting, 2, 32, 70
Load shifting, 40
Low level noise, 68–82
Lymphocytes, 6, 9, 10

M

Management, 90
Manchester Board of Health, 1
Manual dexterity, 32
Marital state, 38
Mean Thermal Vote, 71, 72
Memory, 29, 54
Mental efficiency, 28–30, 69
    effects (Trichlorethylene), 50–55
    health, 83–91
    load (noise), 79
Mesothelioma, 49
Metatarsals, 40
    guards, 41
Metatarsalgia, 39
Micturition (frequency), 65
Migraine, 60
Military situations, 29
Mirror-drawing, 29
Moistness, 40
Motivation, 28

Moving, 88
Multiplication, 29
Muscular tension, 79
"Music", 65

N

Nails, 39, 44
Narcosis, 65
Narcotic, 50
Nausea, 52, 89
Nervous system, 9, 14, 54, 55
Neurasthenic syndrome, 51
Neurosis, 83, 88
Neutral zone, 69
Neutrophils, 6, 9, 10
Night shift, 28–30, 46
Noise, 2, 5, 6, 7, 16, 23, 46, 57–64, 70, 72, 79–81
    accidents, 25
    aircraft, 6
    animal breeding, 25
    imposed, 68–82
    induced hearing loss, 16–24
    maximum area, 6
    pink, 18
    pollution level, 7
    social class, 25
    typewriter, 6
    white, 6, 8, 10, 11
Non-smokers, 58
Nonsense-syllables, 29
Norms (of work), 87

O

Oedema, 41
Optimum temperature, 76–77
Osteo-arthritis, 41
Overweight, 39–40

P

Palpitations, 52
Parameters (stress), 74
Perception, 57, 63, 74
Performance, 28–30, 32, 46, 57, 63, 67, 68, 70, 77

Periostitis, 42
Peripheral vessels, 40
Personality, 5, 54, 72, 80, 85
Perspiration, 40–41
Phenobarbitone, 57–64
Phonetically-balanced monosyllabic words, 19–20
Phonomes, 22–23
Physical activity, 37, 68
    hazards, 83
Piece work, 33
Placebo, 57–64
Plastic, 40
Plantar fascia, 43
Plural noun underlining test, 63
Postural changes, 76
Predicted four-hour sweat rate (P4SR), 71
Pressure, 68
Productivity (and health), 31–38
Psychological illness, 83
Pulmonary complaints, 33
Pulse rate, 58, 60

R

Reaction time, 29
Reading-speed and comprehension, 75, 77, 78
Records, 46
Renal damage, 65
Respiratory rate, 58, 60
Rubber footwear, 40–41, 43

S

Scrotum, 65
Semi-reverberant room, 17
Sensori-motor coordination, 29
Shift-systems, 47
    work, 5, 28–30, 65
Shipwrights, 20
Showers, 41, 44
Sibs, 36
Sickness, 31, 34, 37, 38, 49, 84
    benefit, 38
Signal-to-noise ratio, 19, 21–23
Sinus arrythmia, 73, 78, 79
Skill, 73

Skin resistance, 6, 8, 9, 11, 14
Sleep-deprivation, 28, 29
    difficulty, 85
Smog, 49
Social factors, 32, 70, 88
Solvents, 65
Sound, 63
    pressure level, 68–82
Speech intelligibility, 16
Splinters, 44
Standing, 39
Strain, 68–82, 88
Stress, 9, 65, 66, 70, 83–91
    combinations, 5
    heat, 68–82
    moderate, 68–82
    parameters, 74
Subjective response, 69
Sulphur dioxide, 49
    trioxide, 49
Surface area, 76
Sweat glands, 40, 52
    rate, 68, 69, 71, 76

T

Tea, 58, 63
Temperature, 2, 28, 40, 47, 71, 74–79
    body, 68
    rising, 79
    variations, 39, 46
Thermal stress indices, 71
Thermoregulation, 71
Time estimation, 29
Time variant factors, 72–74
Tiredness, 14
Tobacco, 52, 58, 63
Toes, 40
Toxic hazards, 5
    materials, 46, 83
Traffic fumes, 63
Travel, 89
    distance, 38
Tribromoethanol, 51
Trichloroacetic acid, 50
Trichlorethanol, 51
Trichlorethylene, 49, 50–55, 65
Tsai-Partington test, 77, 79

U

Ulceration, 41, 43

V

Valgus foot, 42, 43
Vapour pressure, 71
Vehicle drivers, 63
Ventilation, 32
Verrucae, 44
Vertigo, 52
Vibration, 68
Vigilance, 29
Vision-monocular, 33
Vomiting, 52, 89

W

Walking, 39
Warmth, 53
Waterproof boots, 41
Weight-lifting, 40
Welders, 65
White cell count, 6, 9, 10
White spirit, 65
Wilson report, 6
Word association, 29
Work efficiency, 1, 68
Working environment, 83–91
    hours, 2
Wood, 40

NO LONGER THE PROPERTY
OF THE
UNIVERSITY OF R.I. LIBRARY